＊

THE SECRET LIFE
OF QUANTA

The
Secret Life
of
Quanta

Dr. M.Y. Han
Foreword by Eugen Merzbacher, Ph.D.

TAB **TAB BOOKS**
Blue Ridge Summit, PA

FIRST EDITION
FIRST PRINTING

Library of Congress Cataloging in Publication Data

Han, M. Y.
    The secret life of quanta / by M.Y. Han.
      p. cm.
    ISBN 0-8306-3397-9
    1. Quantum theory. 2. High technology. I. Title.
  QC174.12.H36 1990
530.1'2—dc20   89-29143
                                      CIP

TAB BOOKS offers software for sale. For information and a catalog,
please contact TAB Software Department, Blue Ridge Summit, PA 17294-0850.

Questions regarding the content of this book should be addressed to:

    Reader Inquiry Branch
    TAB BOOKS
    Blue Ridge Summit, PA 17294-0214

Acquisitions Editor: Roland S. Phelps
Book Editor: David M. Gauthier
Production: Katherine Brown
Book Design: Jaclyn J. Boone
Cover Design and Illustration: Greg Schooley

This book is dedicated to
Bino, Grace, Chris, and Tono

Contents

＊

Foreword

IT IS DIFFICULT THESE DAYS not to be depressed by the growing evidence that the vast majority of our citizens is woefully and dangerously uninformed about science. Whenever I meet someone, I have come to dread the question "And what do you do?" My reply, "I am a physicist," almost invariably draws a response like "Oh, that's a subject I stayed away from in college," or "I have never been any good at math and science." Every survey and test show us that we are faced with an educational problem of colossal proportions. Basic scientific literacy for all citizens must be an essential goal if we and our children are to participate productively in the evolution of our society, our economy, and our culture. A grasp of the elementary ideas and hard-core facts of the physical sciences is indispensable for anyone hoping to lead a full life.

This is not the place for an examination of the causes of the deplorable deficiencies in our entire educational system, which are responsible for fear as well as ignorance of science. If fingers were pointed, the scientific community would not escape blame. Fortunately, the gravity of the problem is being recognized, and change is in the wind. Major improvements of our educational enterprise will be costly in time and money, but there is no reason to wait for the millenium. By and large, newspapers and broadcasting media have begun to shoulder their responsibility by augmenting and improving their scientific reporting, and some serious science books have risen to the top of the bestseller list.

In the physical sciences, many of these popular accounts have emphasized the "weirdness" of the more esoteric, and sometimes speculative, conclusions that we owe to twentieth-century physics and astronomy. The book M. Y. Han has written is more down-to-earth. It starts from the assumption that a gifted expositor can explain to nonscientists how the laws of quantum physics operate in high-technology devices and processes, with which we have almost daily contact. With this program in mind, he has focused on electrons and atoms, which are at the bottom of everything around us. An understanding of the fundamental concepts of atomic structure and of the physical nature of the chemical bond that holds atoms in molecules together provides the tools for answering other questions: How matter in the solid state behaves; how semiconductors, superconductors, and lasers work; and—in the end, by analogy—even what is essential about nuclei and "elementary" particles.

The author is well qualified for this task, which grew out of some lectures he gave in the continuing education program at Duke University. He is a high-energy physicist who has made important contributions to the theory of quarks as constituents of matter. In his teaching he has shown that he knows how to bridge the gap between the arcane subtleties of fundamental particles and the concepts of physics that are within the reach of everyone who has a bit of native curiosity about the environment in which we live. May his efforts bring us closer to a scientifically literate society!

Eugen Merzbacher, Ph.D.
Kenan Professor of Physics
University of North Carolina at Chapel Hill
January 1990, Chapel Hill, NC

Introduction

THE USE OF PERSONAL COMPUTERS has become such a common-place event, from the trading floor of a Wall Street megabank to a neighborhood hardware store, that one wonders how we had ever managed before without them. It wasn't a long time ago either when the PCs came into being. It was only in 1976 that two inventors, Stephen Wozniak and Steven Jobs, hatched the first Apple I personal computer, the predecessor of powerful workstations of today. Similar remarks can also be made with regard to the precise medical operations in which a beam of laser light is utilized as a surgical tool or relatively inexpensive techniques by which genetically-engineered pharmaceutical products are mass-produced.

In a relatively short period of time, the high tech, as advanced technologies of today are collectively called, has achieved very rapid progress all across the industrial landscape and in so doing has brought about a wide range of significant benefits to us as well as profound changes in the way we live and work. It affects our priorities in education, job training, and types of employment as well as our financial well-being. What are the high-tech job trends and investment opportunities between now and the year 2001? What about the new generation of high-tech products ready to be sprung on us? What will it be like going into the 21st century?

We are also affected by larger issues involving high-tech research and development that have far-reaching political as well as economic consequences:

- New generations of computer superchips and their impact on international trades.
- Supercomputer proliferation and the national security.
- The strategic defense initiative.
- The megabillion-dollar global TV market and the impact on it by the high-definition TVs.
- Superhot competition in the race toward commercialization of the high-temperature superconductors.
- The pros and cons of multibillion dollar 'big science' projects such as the superconducting super collider.

In order to be able to make informed decisions, it is highly desirable, if not downright necessary, for the public at large to have an opportunity to gain basic knowledge of what the high-tech is all about. This includes government personnel, politicians, teachers, school administrators, corporate executives, investors, money managers and financial analysts—not to overlook word-processing poets. There is just one problem however.

All high-tech products have been created on the basis of our detailed knowledge of atomic structures and we have gained this knowledge on the basis of theoretical foundations of quantum mechanics, one branch of modern physics that deals with strange behavior of the microcosmos of atoms. The trouble is that quantum mechanics is a subject that is difficult to understand not only for laymen but also for professionals. It is mathematically complex, and its concepts are difficult to come to terms with; as a result, there are few books on this subject available to laymen. Even some rudimentary understanding of quantum physics represents an almost totally impenetrable barrier for laymen to overcome and this situation has effectively denied the public at large an opportunity for an essential understanding of high-tech products and their wonders.

I have adopted what I believe to be a new approach, rarely if ever done in a general science book. Insofar as the results of quantum mechanics have been firmly established for well over a half century by now, why not simply jump over this barrier of quantum theory, instead of trying to burrow through it, explain in a simple language the structures of atoms, and thereby provide a basis for qualitative understanding of high-tech to as many readers as possible? That is the major thrust of this book. After preliminary excursions in the first three chapters, detailed energy level structures and the electron configurations of many-electron atoms are laid out in the fourth chapter.

This is a subject matter usually reserved toward the end of a modern physics course. As you will see, there is nothing really all that difficult about it, as long as you know how to label and count things; this time a whole bunch of electrons.

Now there exists an interesting correlation, an unmistakably inverse correlation, between the amount of texts devoted to quantum mechanics in a modern physics textbook and the number of popular science books available to laymen on the same subject. Typically, no less than 75 percent of a textbook in modern physics is consumed by the exposition of quantum mechanics, the rest of the text is divided about evenly between the subjects of space, time, and of the world of subnuclear particles. On the contrary, no less than 75 percent of general science books deal with the subject of space-time, the theory of relativity and cosmology. Relatively very few popular books are available, no more than two dozen at most, on the subject of quantum mechanics. A rather convincing testimony to the difficulty of the subject.

More often than not, books intended for a general readership end up falling into one of two categories. First, there are the "physicists-to-physicists only" popular books, the kind that can be understood by participants to a special seminar, then there are the "physicists-to-other scientists" popular books. Other scientists with advanced degrees, that is. If this book should prove to be of some help to those uninitiated in quantum physics in gaining qualitative understanding of how atoms form molecules, solids, and doped semiconductors as well as how atoms and light trade energies to produce a beam of laser light, then the new approach that I have taken will have served a purpose.

1

＊

High-tech:
technologies
of the atomic kind

IT BEGAN TO GRADUALLY DEVELOP soon after the end of World War II, coming into a full bloom during the 1970s. By the '80s, it constituted a significant portion of the global economic output—a rapid-fire advancement of highly sophisticated technologies the likes of which did not exist barely one generation ago.

- Personal computers
- Powerful desktop workstations
- Facsimile machines and laser printers
- Precision laser surgeries and satellite communications
- Superconducting magnets and high-energy particle accelerators
- Recombinant DNA technology drugs and magnetic resonance imaging diagnostics

The high technology has spawned a wide range of new industries representing the cutting edges in research, development, and manufacturing. What at first appears to be diverse industries are, in fact, rather closely interrelated. Robotics, supercomputers, and artificial intelligence are direct extensions of the semiconductor and computer technologies, while the knowledge of electronics and laser optics is brought to bear on telecommunication industry. Aerospace and new materials science represent integrated efforts of nearly every branch of high technology. Today's high-tech industries are comprised of the following major branches:

High-tech industries	Fiber optics
Semiconductors	Telecommunications
Computers	Superconductors
Robotics	Aerospace
Supercomputers	New materials
Artificial intelligence	Particle accelerators
Lasers	Biotechnology

High-tech medicine

Underlying the new industries are four basic areas of high technology: semiconductors, lasers, superconductors, and genetic engineering. A brief study of the chronology of important developments in each of the four areas brings out one common denominator. With one or two exceptions, they all happened in the second half of this century. All major events can be traced to their beginnings in the fifties, picking up steam through the sixties, and leading up to the explosive growths we are witnessing today. Let us have a look.

The high technology of semiconductors encompasses a very wide range of products that includes transistors, integrated circuits, memory chips and microprocessors, computer hardware and software, personal computers, supercomputers, robots, and artificial intelligence. Some of its landmark events are as follows:

1947   The transistor was invented by John Bardeen, Walter Brattain, and William Shockley, who were awarded the 1956 Nobel prize in physics for this work. The name *transistor* originally was a contraction of *trans*fer re*sistor.*

1959   A new technique was invented in which a transistor can be manufactured by constructing layers of semiconducting materials. This type of transistor is called a *planar transistor.* This technology opened the way for a mass production of transistors from flat silicon wafers.

1961   A truly landmark year. A complete self-functioning electronic circuit consisting of scores of parts was manufactured as a single piece out of a silicon wafer. This is the integrated circuit (IC), the chip, which was to usher in the age of information. The credit for its invention is shared, independently, by Jack Kilby and Robert Noyes. The IC gets abbreviated to a single letter *I* when combined with other words, such as in an LSI, a large-scale integrated circuit, or a VLSI, a very large scale integrated circuit.

1971 The first microprocessor, a complete computer on a single chip, became a reality. The chip, known as the Intel 4004 chip, contained about 2,000 components. A four-megabit chip of today contains, on a chip about the size of a baby's fingertip, roughly 8 million parts.

1977 The Apple II personal computer, with all the memory power of four kilobytes, appeared and set in motion what we now call the PC revolution. If you are not comfortable with computer terminologies, they will be explained in chapters 6 and 7.

1980 In this year the microcomputer revolution began in earnest, followed by the introductions of the IBM PC in '81, IBM-PC XT (extended) in '83, Apple MacIntoshes in '84, IBM-PC AT (advanced technology) in '86, and IBM-PS/2 in '87.

The high technology of lasers has given us a lot more than laser printers and compact disks. It has spawned an entirely new field of industry, called *optoelectronics,* which includes the fiberoptics telecommunication, as well as CD data storage systems for computers. The technology is also the anchor for the strategic defense initiative (SDI) weapon system. Its history closely parallels that of semiconductors.

1954 James Gordon, H. J. Zeiger, and Charles Townes invented the first operational maser, a microwave counterpart of the laser.

1960 The year in which the first laser was successfully tested. The first laser was a solid laser using a ruby rod, giving off a beam of red laser light, the same kind you see at a grocery checkout counter.

1972 The speed of light was measured to an unprecedented accuracy using a laser beam produced by a helium-neon gas mixture. For those who cherish accuracy to the very last digit, the speed of light was measured to be $299,792.4562 \pm 0.0011$ kilometers per second.

1980 Applications of the laser technology went into a full swing. Industrial applications include cutting, welding, aligning, and surveying. It plays a crucial role in the field of precision surgery, as well as in optoelectronic telecommunication, not to mention the laser guidance system for missiles and bombs.

The high technology of superconductors has a history that is both old and new. The phenomenon of superconductivity is old, having been first discovered way back in 1911 by Heike Kamerlingh-Onnes, but the whole branch of physics remained in a relative obscurity for well over 70 years until a bombshell of a discovery was made in the fall of 1986. Now everyone is in a frantic dash to get into the new high-temperature superconductors with a sideshow of multi-million dollar lawsuits over the patent rights. The dream of superconductor research and development includes inexpensive and highly efficient power transmission as well as power storage, magnetically levitated trains, and superfast switching mechanisms for computer applications. Three recent landmark events are as follows.

1973   The discovery of the first superconducting material, a niobium compound, to display its property above the boiling point of liquid hydrogen.

1986   The discovery of the first superconductor, a nonmetal, that begins to superconduct at temperatures above the boiling point of liquid helium.

1987   The discovery of the first material that superconducts at above the boiling point of the relatively inexpensive and readily available liquid nitrogen.

The discovery of 1986 won the Nobel physics prize for Alex Müller and J. Bednorz. The 1987 discovery kicked off a worldwide mad rush.

The high technology of genetic engineering is the only high technology in life science, and it covers a wide application in the fields of the DNA molecules, recombinations of manipulations of DNAs, and the resulting engineering in agricultural, petrochemical, and pharmaceutical biotechnology. This is one area of high technology that lies outside the scope of this book, but a short chronology is included for the sake of completeness.

1953   The structure of a DNA molecule was discovered to have the shape of a double helix by James Watson and Francis Crick.

1961   The basic code of life was cracked, connecting the sequences of base molecules on a DNA molecule to those of amino acids on a protein molecule.

1978   The first genetically engineered drug was mass-produced: human insulin.

1986   The first hepatitis vaccine produced by the DNA recombinant technology.

1988   The first patent ever given for a genetically engineered living animal, a cancer-prone mouse.

Life science clearly lies outside the scope of physics, and we will not have a discussion of genetic engineering in this book, except a brief encounter in chapter 5 with molecules of life and examples of a specific type of interatomic bond known as the *hydrogen bond.*

Being the products of the second half of this century is not the only common thread running through the four basic areas of high technology. They have one fundamental common denominator at a deeper level: the common source of knowledge, none other than the detailed knowledge of atomic structures unveiled to us by theoretical foundation of quantum physics, all developed during the first half of the twentieth century. Semiconductors and superconductors owe their properties to the electron structure of atoms making up the material, while a beam of laser light is produced by a chain reaction of atoms making transitions from higher to lower energy levels. The mechanism by which two strands of a DNA molecule split and replicate, the very process of life, depends on the structure of atoms and molecules. If a single and universal definition is to be sought for high technology, it must be that it is the technology of quantum physics, the technology indeed of the atomic kind. In the next four chapters, we will go straight to the heart of the matter—the electron structure of atoms.

2

✳

Atoms: building blocks of the universe

ONE OF THE AMAZING THINGS about the universe is that all types of matter in it, living as well as nonliving, are made up out of a basic set of atoms, a finite number of them. From one end of the universe to the other, for all their infinite varieties of forms, all things are basically different combinations of some 100-odd different species of atoms.

Basic properties of atoms

Strictly speaking, atoms are not the ultimate building blocks of matter. They have a layered structure of their own: progressively smaller structures of subatomic and subnuclear worlds. But for all things larger than the subnuclear world, an atom consists essentially of two constituents: the central nucleus, and a group of electrons orbiting around it.

Atoms combine to form molecules. Some molecules contain only a few atoms such as those of oxygen, water, or nitrogen, while others such as polymers correspond to strings of a few hundred atoms each, and still others have even tens of thousands, as in DNA genes. Somewhere along the process of forming two strands, twisted up in the shape of a double helix, out of inanimate nitrogen compounds called bases, life appears within an aggregate of molecules. Other molecules form crystals, liquids, and solids, and they in turn form rocks, oceans, and earth, and so on.

Atoms are to the universe what letters of an alphabet are to a language, say English. Alphabets combine to form words, the "molecules" of English. Some words contain only a few letters such as *it,* a two-letter word or a "two-atom molecule," while others consist of as many as 27 letters. The words form sentences, paragraphs, and chapters, and they in turn form books, libraries, and so on. Even a book on high-tech physics cannot escape this process of formation!

Just as alphabet letters have basic characteristics, all atoms share a set of basic defining properties:

- Exceedingly small in size
- Mostly naturally stable
- Electrically neutral
- Contain energies in discreet amounts
- Absorb and emit light
- Form molecules
- Account for all chemical, physical, electrical, magnetic, thermal, and optical properties of matter

Before we examine how quantum physics provides explanations for these basic properties, as well as other aspects of atoms, there is one item we should settle first: the question of just how many different elements there are. If someone should ask how many letters there are in the alphabet, the answer is simple. There are exactly 52, the uppercases and lowercases each of 26. How many numerals in the decimal system? Precisely ten. Now for something as fundamental as elements or atoms, the building blocks of the universe, you would think that the number of species should be precisely exact. Right? Wrong.

Until the early fifties, the number of elements was 92. That number is still the total number of elements that exist in nature. Since then physicists have managed to push this number steadily upward by either discovering heavier atoms as the byproducts of nuclear reactions or actually producing artificial atoms by high-energy collisions between other nuclei. By 1961 the number had climbed to 103. Some books now list up to 105, and it is currently believed that there are up to 109 different species of atoms.

Two key words to remember here are *naturally* and *stable.* Most of the elements that we are familiar with in our everyday life are both stable and occur naturally, but some atoms occur naturally but are unstable—the naturally radioactive ones—and others are neither stable nor found in nature—the artificially produced elements. Only

81 elements are naturally stable, and 11 elements are found in nature but unstable. Beyond these first 92 the atoms are all artificial and have very short lifetimes.

The limit on the number of stable atoms comes from the limit imposed on the size of the atomic nucleus by the conflicting nature of the electric and strong nuclear forces that act among protons and neutrons inside a nucleus. This point will be discussed in detail later in chapter 11. It is totally acceptable to say that there are, in the whole universe, 103 different species of atoms, about twice the number of alphabet letters, and allow for a handful of very short-living and laboratory-produced exotic species.

Electrons: tiny, versatile, and forever.

The name *electron* owes its origin to the Greek word, *elektron,* for amber, a brownish-yellow fossil resin often used to make pipe stems and assorted jewelry accessories. When rubbed with a piece of cloth, an amber attracts dust and small pieces of paper—the property we know as static electricity.

For a long time, the phenomena of electricity and magnetism were considered to be entirely separate and different from matter. When a piece of a wire became electric, no amount of matter was apparently added to or subtracted from it. Only in 1897, four years shy of the twentieth century, was it discovered by J. J. Thompson that an electricity was actually carried by a motion of extremely tiny bits of matter, since named *electrons.* Not only are electrons the tiniest carriers of electricity, but in fact, the amount of electric charge residing on a single electron defines a basic unit of electric charge for all other known particles observed to date. If some elementary particles are more elementary than others, then the electron is certainly one of the most fundamental.

All of us have a more or less definite mental picture of an atom. We ought to. We have been exposed to schematic drawings of the atom for a number of years, and it is sort of drilled into our minds— a dot at the center of several concentric circles and a bunch of smaller dots, electrons, circling in orbits, just like our Solar System. Before we deal with this picture of atoms, remember that electrons can and in fact do exist quite well just by themselves.

An electron gun shoots out a beam of electrons, which strikes a special coating as it scans the interior wall of a picture tube, producing the images that we see on televisions and computer monitors. A sudden decelerating beam of electrons gives us the soft diagnostic X rays. In a special class of high-energy accelerators, called the *electron-positron colliders,* one beam of electrons and another beam of *positrons,* or anti-electrons, are accelerated in opposite directions to speeds up to 99.99995% of the speed of light and then guided into each other's path for collisions, recreating in a minute scale what the Big Bang must have been like. At present five such electron-positron colliders are operating in the world: two in the U.S., and one each in the U.S.S.R., Japan, and Germany. Two more are nearing completion: one small one in China, and the world's largest in Switzerland.

An electron is an exceedingly small bit of matter. So far as can be measured, it has practically no discernible size, being smaller than 10^{-18} meters; that is, one one-millionth of one-trillionth of a meter. A beam of very fast high-energy electrons is hurled at target electrons, and the shortest distance it can approach before getting deflected by their mutual forces sets an upper limit to the size of a particle carrying an electric charge. An electron is the closest real thing to the abstract physical concept we so dearly cherish: a *point particle*—a sizeless geometrical point having a definite mass, and one of the conceptual pillars of all classical physics.

What is small or large is, of course, relative. A meter is a convenient human-sized unit for length, and an electron is incredibly small in terms of it. The size of an electron is to the size of a human being what a human being is to the size of the Milky Way galaxy. That sure helps a lot, doesn't it?

When dealing with very small or large numbers, the standard scientific convention is to use the notation of exponents, the powers of ten, instead of carrying along a string of 10, 20, or sometimes even 40 zeroes. This method will be used throughout this book. The standard prefixes for powers of ten are summarized in TABLE 2–1. An electron weighs in with a mass of about 10^{-30} kilograms—1 kilogram is equal to 2.2 pounds or 5,000 carats. About 50 trillion trillion electron would weigh about one-quarter of a carat (as distinct from *karat*).

Let us try another analogy. The mass of an electron is to the mass of a mosquito what the latter is to the mass of our Sun. We can try all sorts of other analogies but the fact is that there are simply too many zeroes involved for us to comprehend the relative sizes truly comfortably. Of all known elementary particles with a finite mass, the electron is the smallest, as well as the lightest.

The principal significance of an electron lies not in its size or mass, but rather in its electric charge. A unit for electric charges is

TABLE 2-1 *Powers of ten*

Prefix	Power	Decimal	Name
tera	10^{12}	1,000,000,000,000	trillion
giga	10^9	1,000,000,000	billion
mega	10^6	1,000,000	million
kilo	10^3	1,000	thousand
milli	10^{-3}	0.001	one-thousandth
micro	10^{-6}	0.000001	one-millionth
nano	10^{-9}	0.000000001	one-billionth
pico	10^{-12}	0.000000000001	one-trillionth

called a *coulomb*. The amount of electric charge of an electron is 1.6×10^{-19} coulombs, and by the convention for the relative signs of charges adopted long before the discovery of electrons, the sign of the charge is taken to be negative. A typical hand-held pocket calculator runs on a 1.5 volt (V) watch battery and has a power consumption of, say, 0.0006 watt (W). Because 1 watt is defined as a unit of power corresponding to 1 volt multiplied by 1 ampere (A), this calculator runs on a current of about 0.0004 A. (One ampere is defined as the passing of 1 coulomb of electric charges across any point in 1 second of time. This power consumption of 0.0006 W equates to about 2.5 trillion electrons passing a point in its circuitry every millisecond. The best-known values for the physical parameters of an electron are as follows:

Size: $< 10^{-18}$ meters
Mass: $(9.109534 \pm 0.000047) \times 10^{-31}$ kilograms
Charge: $-(1.602189 \pm 0.0000046) \times 10^{-19}$ coulombs

The magnitude of electric charge on an electron defines the basic unit of electric charge for all particles, ionized atoms, and molecules. No deviation from this rule, the rule of *charge quantization,* has ever been observed. The value, not the negative sign, of an electron charge is denoted by a letter q; all *known* particles can have electric charges only in whole number multiples of q: $+ 3q$, $+ q$, $- 2q$, and so on. An electron has a charge of $-q$ in this notation.

You have heard about *quarks,* which are supposed to make up protons and neutrons. It has been theorized that quarks carry electric charges in the amount of $+ 2/3$ q and $- 1/3$ q. We have talked so much about these quarks for so long, over a quarter of a century now, that sometimes we tend to overlook the fact that they remain unobserved to date.

Another basic property of an electron is its stability. It is absolutely stable, a distinction shared by only a handful of other elementary particles. They do not transform or transmute themselves into other smaller bits under any circumstances. They are the lightest among all particles with nonzero masses. An electron is, therefore, one of the oldest, if not *the* oldest, particle. From the very moment of creation till the end of time, it has been, is and will be around. Diamonds are not forever, but electrons are.

The nucleus: the atomic "sun"

The discovery of electrons and the subsequent realization that they are contained in all atoms gave us the first clue to solving the puzzle of the atomic structure. Because all atoms in their natural state are electrically neutral, some constituents other than electrons must exactly cancel out the negative charges of electrons. This cancellation is very precise, not approximate in any way, to the very last known digit. These other constituents must be responsible for almost all of the atomic mass because the mass of electrons compared to that of atoms is negligible. In a hydrogen atom, slightly less than 0.1% of the atomic mass can be accounted for by an electron.

The fact that the structure of atoms is quite similar to that of a planetary system was established through experiments by Earnest Rutherford during the period 1911 through 1913. In this so-called nuclear model of atoms, almost all the mass of an atom is concentrated in a pointlike object called the atomic *nucleus* (the name was borrowed from the term *cell nucleus* in biology), which carries the exact amount of positive charges to render an atom electrically neutral. The nucleus acts as the center of attractive electric forces, keeping electrons in orbit around it much the way planets revolve around a central star under the influence of gravitational forces.

An atomic nucleus is an aggregate of particles called protons and neutrons that are closely packed together, somewhat like a bunch of grapes. A *proton* is another name for the nucleus of a hydrogen atom, the simplest atom in the universe and made up of one proton with one electron revolving around it. If you were to take a bottle filled with hydrogen gas and subject it to the strong influence of an electric force, you would remove electrons from hydrogen atoms and be left with a "gas" of protons. The size of a proton is determined in the same way as the size of an electron is: by the farthest extension of its electric charge distributions as can be checked experimentally. It is about 2×10^{-15} meters across. Using the prefixes given in TABLE 2–1, it is either two micronanometers or two millipicometers! Actually, there is another standard unit, called a *fermi,* defined as 1×10^{-15} meters. A proton is two fermis long.

An electron has almost no discernible size to speak of—less than 10^{-18} meters—but a proton is about 1,000 times larger than an electron, perhaps even larger. Still, a string of about 11 trillion protons would only equal 1 inch in length. The mass of a proton is about 1,900 times heavier than that of an electron, but when it comes to comparing their respective charges the two are on an equal footing. The electric charge of a proton is identical in magnitude but opposite in sign to that of an electron—such disparities in sizes and masses, but a perfect balancing act in electric charges. The best-known values for a proton are as follows:

Size: 2×10^{-15} meters across
Mass: 1.672×10^{-27} kilograms
Charge: $+ (1.602189 \pm 0.0000046) \times 10^{-19}$ coulombs

A neutron, first discovered by James Chadwick in 1932, is an electrically neutral close cousin of a proton, and except for contributing to atomic masses it plays no role in atomic physics. Instead, the arena for the importance of neutrons is the study of subnuclear structure. We won't be discussing neutrons until we get to chapter 11.

The parameters of a neutron are as follows:

Size: 2×10^{-15} meters across
Mass: 1.675×10^{-27} kilograms
Charge: 0

A neutron is just a tad heavier than a proton. From the time a neutron was first discovered, this point has been the source of many speculations and theories, but to this date no convincing argument exists for this little bit of difference in masses.

Although electrons are truly elementary—not being combined from anything else—protons and neutrons are considered to be composite objects. For a little over 25 years now, a great body of experimental data has been accumulated indicating, indirectly and circumstantially, that protons and neutrons are made of smaller constituents, called *quarks*. In spite of this composite nature, protons seem to share with electrons one of the priviledged properties of being absolutely stable, not transforming or transmuting themselves into other lighter particles.

This stability of a proton has been recently called into question by some theoretical attempts to unify all basic forces. To date, however, no experimental data have been discovered to support such instability, and until such evidence can be found protons are still thought to be absolutely stable. The possible mathematical instability of a proton, even if it is true, is in the order of 10^{35} years, and you need not worry about everything going up in smoke tomorrow.

A perspective of an atom

The way in which we draw a picture of an atom has become so standardized that sometimes it is a little difficult to shake it off and modify it in our mind. An atomic nucleus, an aggregate of protons and neutrons, is placed at the center of several concentric circles, which represent the orbits of revolving electrons. What keeps them all together is the attractive electric force between the electrons and the protons. It is indeed very much like a picture of a planetary system. The picture, as shown in FIG. 2–1. is basically not incorrect, but just a little too neat for an object that exists in the quantum world. It might sound funny, but the usual picture needs to be fuzzied up a bit to convey a more realistic picture.

First, let us get a better handle on the proportionality of atomic and nuclear sizes. Nuclear sizes range from about 2 fermis across for a proton up to about 13 fermis for a uranium nucleus. One fermi corresponds to a length of 10^{-15} meters. Just as a fermi signifies a typical nuclear scale, atomic sizes are often expressed in terms of a unit called an *angstrom,* defined as being equal to 10^{-10} meters. One angstrom, or $\frac{1}{10}$ nanometer, is equal to 100,000 fermis. Sizes of all atoms are fairly constant, varying within a narrow range of 1 and 3 angstroms.

Let us compare the relative sizes of an atom as a whole and its nucleus, taking as an example a carbon atom, an atom of some importance to us:

Atomic size: 3 angstroms, 3×10^{-10} meters
Nuclear size: 5 fermis, 5×10^{-15} meters

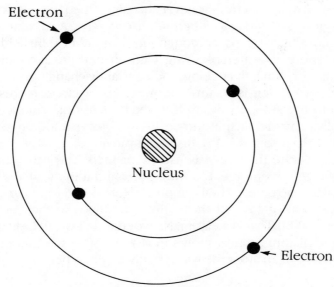

2–1   *A familiar picture of an atom*

The ratio of the two sizes is 300,000: 5, or 60,000:1. It is impossible to give a realistic representation of the 60,000:1 ratio to any picture within the bounds of a page in a book, or the size of a standard blackboard for that matter, and still make the dot at the center, the atomic nucleus, visible!

Let us try another approach. If we blow up the carbon nucleus to the size of 1 foot, the carbon atom would be 60,000 feet across; that is, 3.8 miles across with a circular circumference of about 12 miles. Imagine an airport whose main runways are about 3 miles long. Right in the middle of a runway in the dead center of the airport, whose boundaries are marked by 12-mile-long chain fences, sits a basketball about 1 foot in diameter. At a far corner of the airport, at the foot of a chain fence, hidden under some weeds, is a pea-sized pebble. Thus, we have the airport, the fences, a pebble, and a basketball. Whenever we see a picture of an atom similar to the one shown in FIG. 2–1, we need to keep this perspective of relative sizes squarely in our mind.

The second point that needs elaboration is more physical than the question of proportions. A sharply drawn circle depicting an electron orbit is too much of an abstraction, and this aspect touches the very conceptual foundation of quantum physics. The position of a particle, such as an electron, is never sharply located but is spread out over a region, giving rise to inherent uncertainties. A circular orbit is actually more like a *torus,* or doughnut, with rather diffuse boundaries. It is more like a smoke ring with an uneven distribution of smoke, that is rolling, fluctuating, and constantly changing.

Now, everything is in three dimensions. A circle represents a sphere and the smoke ring, well, three-dimensional smoke shells, one wholly contained inside another! Try to blow some smoke shells! So we come to a somewhat diffused-out picture of an atom, which is far more accurate than what is usually painted. The heavy nucleus defines the center of force. Electrons surround it in the shape of layers of concentric diffuse shells, appropriately called *electron clouds.* The proportion of sizes of these electron clouds to the nucleus is typically 60,000:1. That is an atom! No wonder that we all hide behind a nice little picture such as shown in FIG. 2–1. Who would and could draw a picture of an atom the way it really ought to be?

Energy levels of the electron shells

The most striking and perhaps the most significant property of an atom is the fact that its energy is quantized; that is, its energy is restricted to a certain discontinuous and intermittent set of values. Energy is essentially a measure of a force, and electrons inside an atom draw their energies from the electric force between the nucleus and themselves. To say that the energy of an atom is quantized is to imply that these electrons can exist only in a certain discrete set of orbits, or shells, that have correspondingly discrete values of energy. It would be like a hypothetical skyscraper in which there are only an intermittent number of floors: 3rd, 8th, 16th, 33rd, and 64th. I do not mean that elevators stop only at these floors, but that those are the only floors in the building and there is nothing in between!

The fact that the energy of atoms is discretely valued is what prompted the invention of quantum physics. It is one of many strange behaviors of matter that can be observed only within the confines of the small world of atoms, and as such has no counterpart in terms of human-sized objects. To get by this kind of difficulty, we will just have to be quite inventive in cooking up some imaginary situations. Such examples abound in any introductory physics textbooks. A block is dragged over a totally frictionless surface pulled by a totally massless rope attached to it. A metal ball is attached to one end of a massless string, which is strung over a frictionless pulley, the pulley itself hung from a ceiling by a massless spring—an infinitely thin wire that is absolutely rigid! We pretend these things are readily available off the shelves of any corner drugstore. In the same vein, let us consider two examples cooked up to display discrete values of energy: a "quantum" spring and a "quantum" satellite.

Consider a coil of a spring. The more it is stretched or compressed, the more energy is stored in it. A simple experiment with a pain-producing excercise machine will convince you of that. The force of the spring, wanting to return to its normal length, is the source of this energy. In the same way, the attractive force between electrons and an atomic nucleus is the source of energy for electrons. To say that an electron can have only a certain set of values for its energy is to say that a coil of a spring can be pulled out or compressed into only a certain set of predetermined extensions. An absurd spring like that deserves to be called a "quantum" spring!

Suppose one end of a spring is attached to a wall and the other to a block, as shown in FIG. 2–2. The block slides across a flat floor—a frictionless floor, of course. Its normal unstretched length is marked as position 1. As you apply a force to it and attempt to pull the block to the right, something totally strange will unfold. As you carefully and gradually increase your pull, absolutely nothing happens; the spring simply does not budge—that is, until you have increased your pull to the exact amount required to extend the spring to the position 2. At that point the spring will suddenly stretch out to position 2. The same pattern will repeat with respect to positions 3, 4, and so on. No other extension of the spring is physically possible, except the allowed set of discrete positions. The energy of the spring at each of the allowed positions is clearly related to the amount of its extension. Energies are labeled as E_1, E_2, E_3, and E_4, and it is obvious that E_4 is greater than E_3 which, in turn, is greater than E_2, the lowest energy being E_1.

The second example, a "quantum" satellite, involves an imaginary situation in which available orbits of an Earth satellite are severely limited (FIG. 2–3). Of all the space that surrounds Earth and all orbits in it that are mathematically possible, this "quantum" satellite cannot exist anywhere except for certain trajectories. This example might not be as easy to visualize as the first example, but in terms of physical properties it is actually much closer to the atomic situation, not surprisingly because gravitational and electric forces have identical geometrical dependence. Both forces vary inversely as the square of distances. We will return to this similarity in more detail in chapter 4.

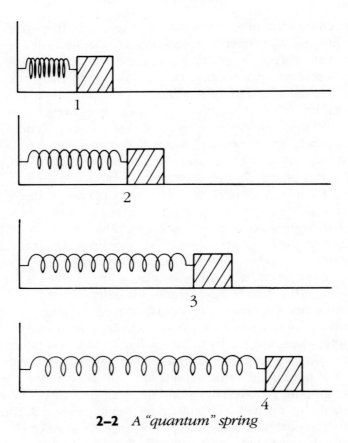

2-2 *A "quantum" spring*

Just as in the case of a spring, energy E_3 is greater than E_2, E_2 is greater than E_1, and so on. It requires more force—that is, more work by a rocket booster—to place a satellite into a higher orbit.

It is much more convenient, and hence is a standard practice, to represent the quantized values for energy of a physical system simply in terms of the set of numbers E_1, E_2, E_3, E_4, E_5, etc., instead of physical descriptions such as the rank of orbits or the order of extensions of a spring. The values of energy are marked on an appropriately chosen vertical scale and graphically represented by short horizontal lines. This diagram is called an *energy-level diagram*. It is an abstract representation of a physical system that is quantized and as such can easily appear to be devoid of meaning to those uninitiated. An energy-level diagram for either examples we discussed, or for an actual atom for that matter, may look like the diagram given in FIG. 2–4.

Every physical system having a discontinuous pattern of energy

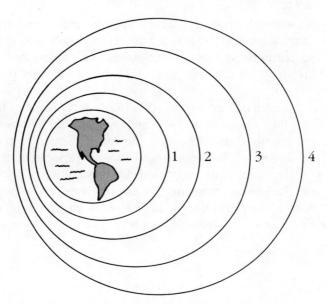

2–3 *A "quantum" satellite*

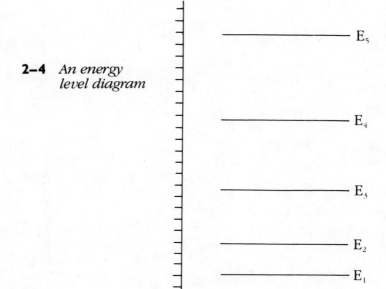

2–4 *An energy level diagram*

can be represented by such a diagram. The spacings between the levels, the energy gaps, and the actual numerical values of energies will be different from one system to another. Each atom has its own characteristic energy-level diagram since each has a different number of electrons.

The two terms *electron shell* and *energy level* refer to the same physical attribute. A shell describes the physical dimensions of an orbit, while an *energy level* represents the value of energy an electron possesses in that shell. The most significant and spectacular triumph of quantum physics was the complete understanding it provided of the electron shell structures and the energy levels of all atoms. We will return to this important topic in chapter 4 after introducing electromagnetic radiation in the next chapter.

3

✳

Let there be light

THE SOLAR SYSTEM is estimated to be about 5 billion years old, give or take a few thousand. Homo sapiens have been hanging around on Earth for some 2 million years, busy doing all sorts of things, one of which was to worship the lifegiver, the Sun, and its light for almost all that time.

Light, electricity, and magnetism

Scientific inquiries into the nature of light did not begin in earnest until about 300 years ago. We came to learn the nature of it, building on the newly gained knowledge of electricity and magnetism during the latter half of the nineteenth century, barely 100 years ago. Electricity, magnetism, and light—the three fundamental aspects of nature absolutely indispensable for our well-being—all share a single common source: the electric charge. Let us dwell upon this point a little.

Sources for the electric force are called *electric charges,* or charges for short. They come in two opposite varieties—positive and negative—and give rise to either attractive or repulsive interaction between unlike and like charges, respectively. The charges, positive or negative, are carried on most, but not all, matter particles. In an atomic scale, as discussed in the previous chapter, the negative charges of an electron and the exactly matching positive charges of a proton serve to define fundamental units for all charges. For reasons not yet understood, electric charges reside only on those matter particles that have some mass, however small. All known particles that have no mass, of which there are a few, are electrically neutral.

In many instances a physical system that is electrically neutral as a whole can be composed of two oppositely charged regions of the same magnitude, separated either temporarily or permanently by a relatively short distance. Such an arrangement is called an *electric dipole*. Electrons of an atom can be pulled to one side by some force external to the atom, shifting the center of negative charges slightly to one side. The positive charges of the nucleus at the center of an atom and the center of negative charges shifted slightly off center would form an electric dipole, as illustrated in FIG. 3–1.

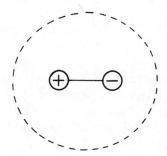

3–1   *An electric dipole*

Many familiar molecules have a permanent electric dipole charge distribution, the so-called *polar molecules*. A water molecule is one polar molecule, the neighborhood of the oxygen atom being a bit negative while the areas occupied by hydrogen atoms being a bit positive. Such familiar but diverse phenomena as friction, viscosity, and surface tension are the results of simple pulling together among polar molecules—the negative end of one polar molecule attracting the positive end of another. Try to explain to a child that large soap bubbles he or she is making in the backyard are held to their fluid shapes by a surface tension resulting from the collective pulling together of billions of polar molecules!

Since both an electric charge and electric dipole have just been mentioned, this may be a good place to make a few brief remarks about the art of terminology. A *charge* refers to a single, individual electric charge, whereas a *dipole* stands for a pair of equal but opposite charges. A dipole is never called a *dicharge* or even a *bicharge*.

Presumably by the same token, an electric charge is never referred to as an electric *monopole*. On the other hand, a magnetic counterpart to an electric charge, mathematically possible but nonexistent, is almost always called a *magnetic monopole*. (Quite often terminologies of science represent the least scientific aspect.)

Sources of magnetic force are not magnetic charges. The reason is simple and direct: magnetic charges do not exist. To be sure, there have been sporadic reports of their sightings, mostly in California, but for well over 60 years since Paul Dirac first raised the possibility of their existence, magnetic monopoles have failed to turn up in all attempts to find them. Instead, the source of magnetism is the *electric charge in motion;* that is, the flow of an electric current. A ring of an electric current, as shown in FIG. 3–2, called a *magnetic dipole,* constitutes the basic unit of magnetism.

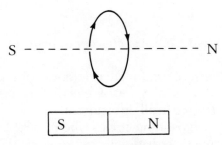

3–2 *A magnetic dipole*

Each side of the ring of current corresponds to the polarity of magnetism, what we are familiar with in terms of a bar magnet as the north and south poles. The familiar bar magnet, one end marked "north" and painted red, and the other marked "south" and painted blue, behaves the way it does owing to a perfect alignment, heads to tails, of a countless number of atoms, each of which is a small magnetic dipole. An atom is full of rings of currents, the spherical shells of electron clouds, remember? FIGURE 3–2 shows that positive charges flow in the direction indicated; the magnetic polarities would be reversed for negative charges in the same direction or for *that matter,* postive charges flowing in the opposite direction. In FIG. 3–3 is summarized the many important roles played by electric charges for electricity and magnetism.

The electric charge also acts as the source for another natural phenomenon, perhaps the most important of all: electromagnetic

	Monopole	Dipole
Electric	$\oplus$ or $\ominus$	$\oplus$ $\ominus$
Magnetic	Does not exist	S $-----$ N

3–3   *Electric charges in action*

radiation. A sudden acceleration, deceleration, or rapid oscillation of electric charges creates a wave of electromagnetic energies that spread out, or *propagate* in all directions in much the same way water waves spread out in expanding circles when a pebble is dropped into a pond. Electromagnetic radiation is also emitted, as will be discussed in later chapters, by atoms and atomic nuclei. These emissions also originate from charges and can be broadly classified into the general category of accelerating charges.

After having concluded, by purely mathematical deductions, the existence of the propagating electromagnetic wave, James Maxwell came upon a remarkable mathematical result. When he calculated the speed of the wave's propagation, it turned out to be identical to the known values of the speed of light! Based on this calculation, Maxwell proclaimed light to be a part of the electromagnetic radiation itself.

In 1887, eight years after the death of Maxwell at the age of 48, the existence of the electromagnetic wave was confirmed by Heinrich Hertz. This is how, barely 100 years ago, we came to learn the true nature of light. Electric charges, whatever they are and however they originated, are more than just fundamental. There is something almost sacred about them. Electromagnetic radiation covers a wide band of the magnitudes of frequencies and energies, only a very narrow portion of which is light. Not only is electromagnetic radiation the lifegiver itself, it is also nature's gift to us for microwave ovens, diagnostic X-rays, and telecommunication.

Electromagnetic radiation

Three different names are normally used to label different portions in the spectrum of electromagnetic radiation: *waves, radiations,* and *rays.* Each description seems to suggest a rather well-differentiated meaning. *Waves* refer to those that are "soft" and harmless such as radio and television waves, shortwaves, and microwaves. Infrared or ultraviolet radiation begins to sound a touch of alarm, while the word *rays* is definitely "hard" and unfriendly—X-rays and gamma rays!

The most striking and singularly important aspect of electromagnetic radiation in general, and light in particular, is the strange fact that its speed is absolute with respect to any relative motion. The numerical value for the speed of light depends on whether it is propagating through a vacuum or a material such as water or plastic. Within a given medium or in a vacuum, however, it has one fixed value, irrespective of any relative motion. Whether an observer is approaching or receding from a light source or whether the light source is approaching or moving away from an observer, the speed of light measured is always exactly the same. You can never, never overtake light. It doesn't even slow down with respect to you, no matter how fast you are chasing after it. This remarkable fact was the only experimental basis that Albert Einstein needed to launch his famous theory of relativity back in 1905.

The speed of light is usually given by its value in a vacuum, which can be expressed in a variety of units as approximately,

3×10^8 meters per second, or
186,000 miles per second, or
1 foot per nanosecond.

More exactly, it is equal to:

2.997924562×10^8 meters per second, or
186,451 miles per second, or
0.9843 feet per nanosecond.

Two principal characteristics of a wave are its wavelength and frequency. *Wavelength* is the distance between two successive crests or troughs of a wave, as shown in FIG. 3–4. It is the length of one complete cycle, one full wiggle, or one periodic repetition, which-ever you prefer. The definition is simple, but the wavelength has a far-reaching significance because it sets the lower limit on the power of resolution for a wave.

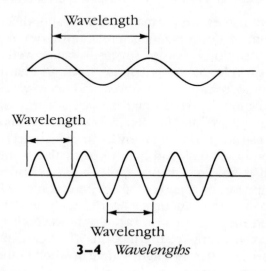

3–4 *Wavelengths*

This question of the resolution power can be illustrated by using a standard 12-inch ruler as an example. One edge of a standard ruler gives the scale in inches, the smallest markings defining $\frac{1}{32}$ inch. It would be a little difficult to ascertain a length of an object smaller than that. Putting it another way, if you are given an unmarked 1-foot ruler, you can measure so many feet by flipping the ruler over and over, but it would be difficult to resolve the difference between $\frac{1}{2}$ and $\frac{1}{4}$ foot with it.

The wavelength of a wave, in the same manner, defines the small-est markings. As the miniaturization processes demand smaller reso-lution in the manufacturing technology of microelectronics, we utilize electromagnetic waves of shorter and shorter wavelength. The so-called *submicron technology* refers to the chip manufacturing technology, in which some electronic components are smaller than a few microns (one micron being another name for one micrometer, one millionth of a meter).

The second attribute of a wave, *frequency,* is the number of com-pleted cycles or full wiggles per given interval of time. Whereas a wavelength refers to a length, a frequency is just a simple number: so many wiggles. The rate of fluctuations of one full cycle per second defines the unit of frequency called a hertz (Hz), named after Heinrich Hertz, who confirmed experimentally the existence of the electromagnetic wave.

After climbing a flight of stairs, if your pulse rate is 120, your heart is pumping at the rate of two pulses per second, a big 2 Hz. The house current comes in at 60 Hz. Some of the fastest personal computers boast a "blinding" speed of 30 megahertz (*MHz*) meaning electric signals are being turned on and off inside the computer at the rate of 30 million pulses per second. The average frequency of the visible light is about 5×10^{14} Hz; that is 500,000 gigahertz (GHz), 500 terahertz (THz), or just plain old 500 trillion oscillations per second.

The speed of a wave can be expressed as a product of its frequency and wavelength. If a wave travels 2 feet in one complete wiggle (the wavelength) and it wiggles 7 times in a second, (the frequency), it is moving at the speed of 14 feet per second. Because an electromagnetic wave knows only one speed, 3×10^8 meters per second in a vacuum the lower the frequency, the longer its wavelength.

FIGURE 3–5 shows the complete spectrum of electromagnetic radiation. Frequencies and wavelengths are paired such that their products are always equal to the speed of light.

As shown in FIG. 3–5, most of the different regions of frequencies are referred to by specific names, but do not have sharp demarkation lines. Ultraviolet rays overlap some portions of soft X-rays region; microwaves and infrared rays have a sizable overlap; and so on. There are two exceptions: the band of visible light and the airwaves controlled by the Federal Communications Commission (FCC). Visible light spans a range of frequencies from 4.5×10^{14} Hz to 7.5×10^{14} Hz, covering the spectrum from red to violet. A red light has a longer wavelength than a blue light, about 700 nanometers for red compared to about 450 nanometers for blue. We will use this fact shortly in discussing the red shift.

Just as the range of visible light is sharply defined by our physiological limitations, the FCC airwave territory is clearly bounded by government legislations. It is already quite crowded and getting worse all the time. In addition to the communication bands listed in FIG. 3–5, we have to contend with cellular phones and marine, aviation, police, and citizens bands! By the early 1990s, the next generation of televisions, the printed picturelike high-definition televisions (HDTVs), will be in full swing and begin to establish a global market worth hundered of billions of dollars. These HDTVs will have to find

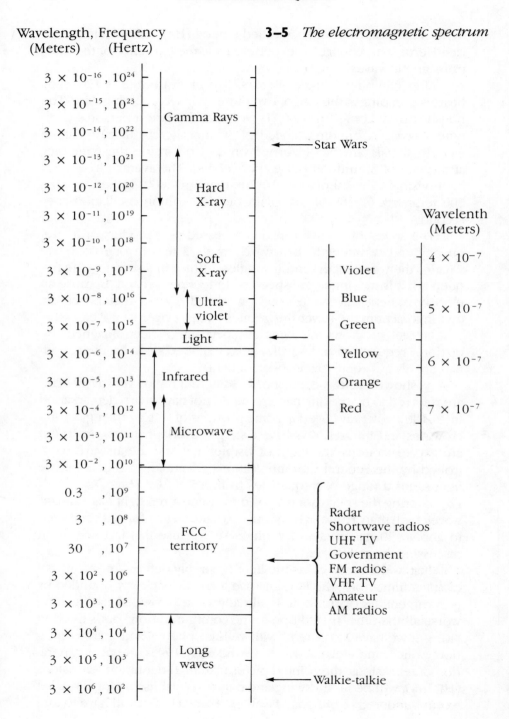

3–5 *The electromagnetic spectrum*

room somewhere within the crowded range of frequencies. Electro-magnetic radiation is the only such radiation of energy in the universe, and that is all we have to play with, crowded or otherwise.

Light and the universe

Almost entire subject matters in this book deal with things that are small, exceedingly small. This section is one of only two exceptions. Here we deal with things that are slightly larger and discuss how the physical nature of light helps us in measuring distances and ages of even the most distant galaxies, all in three simple steps. Each time we discover yet another farthest galaxy, the size of the known part of the universe becomes that much larger. The three steps involve one empirical law and two properties of light: Hubble's law, the Doppler shift, and the speed of light, respectively.

The Doppler shift in sound waves is something we experience almost everyday. As an ambulance or a fire engine passes us on a boulevard or whenever a jetliner flies past us directly overhead near an airport, we hear a higher pitch when it is approaching us and a lower pitch when it is moving away. This shifting of frequencies in relation to the relative motion between an observer and a source of a wave is called the *Doppler shift,* and it provides us with a very convenient and reliable method of determining the speed of a moving object.

The Doppler shift affects the whole range of frequencies in the electromagnetic spectrum, even though it is a little difficult to actually experience this effect with light. When a light source is approaching us, its frequency shifts up, or equivalently its wavelength becomes shorter, and conversely when the source is pulling away from us, the frequency shifts down, or equivalently its wavelength becomes longer. If you wish to personally experience the Doppler shift of the red signal appearing green, you would have to be approaching that traffic light at a speed of about 60,000 miles per second, or about 3.6 million miles per hour, which is about 30% of the speed of light!

On the other hand, the wavelength of light from a source that is moving away from us will become longer, which is to say that it is shifting toward the red end of the visible light spectrum. This is what is referred to as the *red shift.*

One of the basic properties of atoms is that they absorb and emit light. As we will discuss in detail in chapter 8 each atom has a unique set of wavelengths it emits according to its own energy-level structure.

By measuring the red shifts of wavelengths emitted by various atoms contained in a star or a galaxy, we can accurately determine the speed with which the object is moving away from us. This speed leads to a fairly good estimate of how far the object is from us when substituted into a relationship between the receding speed and the distance of a cosmic object. This estimate is known as Hubble's law.

In 1929, Edwin Hubble discovered a remarkable empirical relationship, which states that the receding speed of a star or a galaxy is directly proportional to its distance from us. The farther it is from us, the faster it is moving away from us; the nearer it is, the slower it is moving. Hubble's law is not a perfect law, but it is remarkable in its simplicity, as well as its implication. This was the beginning of our concept of the expanding universe, barely 60 years ago.

With respect to some relatively nearby galaxies, Hubble's law does not apply all that well, and because of uncertainties involved in the determination of great distances by geometric means, the law is usually applied with up to a 20% margin of error added. Nevertheless, Hubble's law applies to a great number of known galaxies and provides us with fairly accurate estimates of their distances from us.

Having obtained the receding speed of a cosmic object by the Doppler red shift and the distance by Hubble's law, we now divide the distance by the speed of light to obtain its youngest possible age. Since no physical object can have a speed greater than that of light, the age of an object obtained this way is its youngest possible age. This is how we obtain the size and age of our universe. Until very recently, the universe was generally thought to be about 15 billion years old. The recent report finding the farthest galaxy to be about 15 billion years old indicates that the age of the universe is still older.

Light and the twentieth century physics

The two-in-one discovery made near the end of the nineteenth century—the existence of electromagnetic radiation and the realization that light is none other than a small portion of it—must be ranked as one of the greatest achievements of all time. The physics of the twentieth century, *modern physics* as it is called, began to be unfolded rapidly soon after. While studying the energy of electromagnetic radiation in a minute scale, Max Planck discovered in 1900 that the energy of light comes in terms of a basic indivisible unit, a *quantum* of light, which was at that time a totally revolutionary concept.

The absolute independence of the speed of light with regard to any relative motion is the foundation from which Albert Einstein launched his famous theory of relativity in 1905. Both branches of modern physics, relativity and quantum physics, trace their genesis to the physical nature of light.

The word *quantum* literally means "the unit of a quantity," the smallest definable amount by which to measure a given physical quantity. By definition, therefore, it stands for something that is discrete and countable. Pennies are the quanta of money. Each grain of sand is a quantum of what otherwise appears to be a continuous expanse of a beautiful beach. If you draw a line with a pencil on a piece of paper, the line looks continuous, but when examined under a

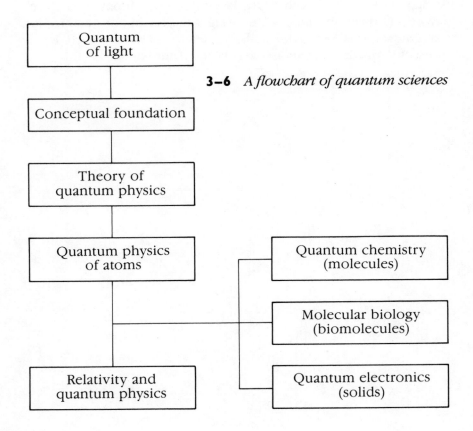

3–6 *A flowchart of quantum sciences*

microscope it is a series of dots marked on a rough texture of the paper.

In physics, *quantum* stands for an elemental unit of energy, a "penny" of an energy for atoms, molecules, and electromagnetic radiation. A flowchart for the development of the science of quanta is given in FIG. 3–6. The first two stages—the conceptual foundation and the formulation of the theory of quantum physics—constitute the standard introduction into the subject, and they were more or less firmly put into their places by the end of 1920s at the hands of a group of all-time greats including Louis de Broglie, Wolfgang Pauli, Max Born, Werner Heisenberg, Erwin Schroedinger, and Paul Dirac. The third stage, the quantum physics of atoms and molecules, serves as the springboard for quantum chemistry dealing with molecules, molecular biology, the study of biomolecules such as DNA, and as the quantum theory of solids and quantum electronics. This is the lifespring of today's high-tech. In the last stage of development, both relativity and quantum physics are brought to bear in our investigation into the secrets of nuclei and elementary particles. Such is the present-day legacy of the discovery of the nature of light.

4

＊

The order
of electrons

THE NAME *ONE-ELECTRON ATOM* pops up quite often and occupies a prominent position in the study of atomic structure. A considerable portion of such a textbook is allocated to a detailed mathematical treatment of a hydrogen atom—the discrete set of allowed orbits, their radii, corresponding energy levels, and so on.

How many one-electron atoms are there?

Composed of one lone electron orbiting around a lone proton, a hydrogen atom is not only the simplest, but also the only true, one-electron system in its natural state—out of some 105 different species of atoms. However, the term *one-electron atom* or *one-electron atom analysis* applies to many other atoms. This point is a bit subtle and needs to be brought out of its relative obscurity.

Even though we have under our command all the mathematical equations and rules that govern, in principle, the behavior of an aggregate of as many charged particles as we like, the actual calculation becomes increasingly complex as we go to larger many-electron atoms. Shell after shell of electron orbits—10, 20, even 70 of them— some spherical, some slightly elliptical, and still others crisscrossing each other, make it a little difficult to keep track of all the interelectron mutual interactions. It turns out that most of the interelectron effects get averaged out and are replaced by a series of successive corrections to a predominant approximation, in which each electron is treated separately, as if it were the only electron. This is one-electron atom analysis. It is a very good approximation technique, and forms the basis of understanding for all many-electron atoms.

In FIG. 4–1 we are shown the two simplest cases of one-electron atoms. The hydrogen atom is, of course, the true one-electron atom, having one electron bound to its nucleus of one proton. The two-electron helium atom is treated, in the one-electron atom analysis, as the sum of two "one-electron" systems, each of which consists of an electron bound to and in orbit around the nucleus with two units of positive charges; that is, two protons. The necessary correction due to the effects of two electrons on each other is then added to these two one-electron systems.

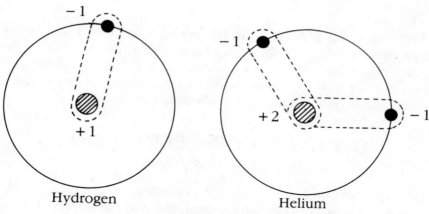

Hydrogen   Helium

4–1 *Hydrogen and helium atoms*

In FIG. 4–2 is shown the case of a three-electron lithium atom. Things get a little more interesting here. According to a set of basic rules, which we will get into shortly in this chapter, the third electron is dispatched to the second orbit, much farther out than the inner, first orbit. Now as far as the two inner electrons are concerned, except for the fact that the nucleus now contains three units of positive charges, the situation is identical to that of a helium atom. As for the third electron, a little reflection will tell you that the situation is right back to that of a hydrogen atom, approximately. The lonesome end sees a central core region whose net charge is +1 unit. It is a one-

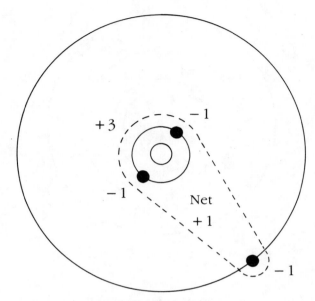

4–2 *A lithium atom*

electron atom all over again. A lithium atom is chemically very similar to a hydrogen atom, for this very simple reason.

Next time you just can't avoid the misfortune of having to engage in small talk with a physicist at a cocktail party, just keep referring to a sodium atom with eleven electrons as a one-electron atom; it might cut short your misfortune.

Orbits of one-electron atoms

Now that we know the approximation procedure, we won't be caught dead talking about a mere hydrogen atom. Instead, we will talk about "one-electron atom" systems. It sounds more professional! Many qualitative features of a one-electron atom can be understood easily by comparing it to a hypothetical example of a "quantum" satellite, discussed briefly in chapter two.

In FIG. 4–3 are shown only the first three orbits for such a satellite. The radii are numbered as r_1, r_2, and r_3, starting from the innermost one. The satellite's energies are denoted by E_1, E_2, and E_3. Keeping in mind that the force in question is the gravitational attraction between the satellite and Earth, let us examine qualitatively how the radii of orbits are related to the energies of the satellite.

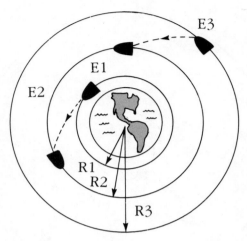

4–3   *"Quantum" satellites revisited*

First of all, the larger an orbit is—that is, the higher it is from the surface of the Earth—the greater is the energy of a satellite. To put a satellite into an orbit, an enormous amount of energy must be expended (the rocket propulsion energy) to overcome the constant gravitational pull. To launch a satellite into a larger, and hence higher, orbit, bigger boosters, more rocket fuel, and a faster speed of climb are needed.

In order to jump from a lower to a higher orbit, a satellite or a space capsule would have to fire on-board rockets to gain more speed, and hence more energy; whereas, to drop down to a lower orbit, it would have to fire retro-rockets to slow down, and lose some energy.

Another aspect of a larger orbit is related to the fact that the gravitational pull is inversely proportional to the square of distance. The farther out an orbit is, the less strongly a satellite is bound to Earth and the easier it becomes for the satellite to escape the gravity of the Earth altogether, and to begin its journey into space.

Needless to say, we know of no electron that is equipped with a retro-rocket! This example of a hypothetical "quantum" satellite is useful only to the extent that it provides us with a convenient mental visualization for a similar situation inside an atom. The similarity between the two should not be surprising because both electric and gravitational forces are inversely proportional to the square of distance. Just as in the case of the satellite, the energies of a planetary electron have the lowest value in the innermost shell, and they increase with the increasing radii.

An exact mathematical solution without any approximation schemes is obtainable for a one-electron atom system, and it yields some interesting patterns. The differences in radii between two successive orbits diverge as we go to larger and larger orbits; that is, the distance between the third and fourth orbits is greater than that between the second and third, and so on. The first four radii are sketched in FIG. 4–4. They are not drawn to any specific scale, but rather spaced appropriately to indicate the pattern of progressive divergence.

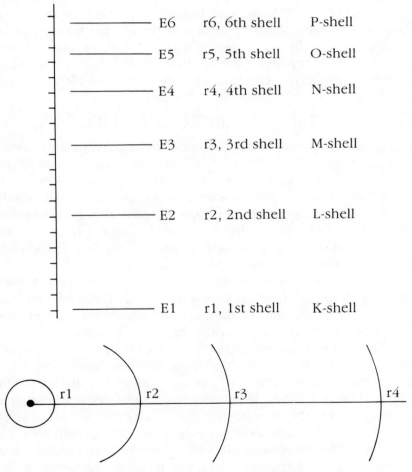

E6	r6, 6th shell	P-shell
E5	r5, 5th shell	O-shell
E4	r4, 4th shell	N-shell
E3	r3, 3rd shell	M-shell
E2	r2, 2nd shell	L-shell
E1	r1, 1st shell	K-shell

r1 r2 r3 r4

4–4 *Energy levels and radii of one-electron atom*

Corresponding energy levels for an electron at each of these possible orbits, or shells, follow an opposite pattern to that of the radii. The differences in values between two successive levels get smaller as the levels get higher; that is, the energy gap between the third and fourth levels is smaller than that between the second and third levels, and so on.

The pattern of progressively narrowing gaps is also sketched in FIG. 4–4. Again, they are not drawn to any particular scale, but only schematically to show the pattern, because we are not particularly concerned with the exact quantitative aspects of these energy levels. The lowest level, with energy E_1, is referred to as the *ground state,* the ground floor of an atom so to speak. Under normal circumstances, an electron always stays in its ground state, the stable configuration whose radius, r_1, serves to define the atomic size. In literature, especially in connection with any study of X-rays emitted by atoms, the energy levels or shells are sometimes denoted by a set of alphabet capitals starting from K, as shown in the figure.

Subshells: clusters of orbits

The pattern of energy levels of a one-electron atom is much richer than the progressive patterns for radii and levels, as described in the previous section. Each shell, or each orbit, actually represents a cluster of highly organized substructures, and what at first appears to be a single energy level consists of several levels closely bunched together. A single horizontal bar, as in FIG. 4–4, representing one possible value of energy an electron can have, is really several horizontal bars spaced closely together.

These substructures are called *subshells.* A shell now is a group name for a family of subshells that are closely bunched as well as related to each other. An equally plausible name, *suborbits,* somehow never made it into the nomenclature of quantum physics.

The hierarchy of energy levels corresponding to subshells of a one-electron atom forms the backbone, the foundation, from which the energy level structure of all atoms is built up, and in this and the next sections we will be delving into it in some detail. Even though the next two sections may be the most difficult sections in this entire book, there are no mathematical equations. The standard system of labeling the family of subshells uses a mixture of numerals, letters, superscripts, and subscripts, and hence it "looks" mathematical. But then a license plate such as 4A6–107X or a nine-digit zip code such as 27715–1254 looks just as mathematical. The use of superscripts will, however, make the labels for subshells look more professional than license plates. We have yet to see a license plate such as MARY-1s² for "Mary is too!"

Still, the next two sections won't be terribly easy either. After all, we are coming to grips with the central results of the quantum theory of atoms. Rarely, if ever, does any popular book on quantum science explain what we are about to do.

The rule that tells us the number of subshells forming a single cluster, a shell, is as precise as it is simple: Each shell is comprised of as many subshells as its numerical rank, starting from the lowest energy level and counting up, as shown in FIG. 4–4. (Five subshells make up the fifth shell; the third shell contains three subshells; and so on.) As we will see in the next section, the first seven shells accommodate all known atoms in the universe, but the rule stipulates, mathematically speaking, that the 100th shell should contain exactly 100 subshells, 1,000th exactly 1,000 subshells, while the 1st shell consists of only 1 subshell: itself. Several clusters of subshells are shown schematically in FIG. 4–5. On the right side of the figure, they are arranged in a telescoped view in order to display the standard labeling scheme used in atomic physics.

The manner in which these levels are labeled is more of an art than a science. Because shells are numbered numerically in an ascending order, you would think that a cluster of subshells within a shell might be labeled in an alphabetical order. It would be only too natural and reasonable to expect a sequence of (1a), (2a, 2b), (3a, 3b, 3c), (4a, 4b, 4c, 4d), and so on. Well, things didn't quite turn out that way. Perhaps someone felt that such a sequence would be too easy and too transparent—totally unbecoming to the quantum theory of atoms. Eventually, it would have looked too much like the classification codes used by the U.S. Selective Service anyway!

The standard names of shells called the *spectroscopic notations* in the trade, start off with a sequence whose first four codes are *s, p, d,* and *f.* Once upon a time, these four letters stood for sharp, principal, diffuse, and fundamental, respectively, but lacking a lasting meaning these words have long since faded out, leaving only the notations. What about after *f ?* After *f,* it becomes alphabetical; that is, *s, p, d, f, g, h, i, k,* and so on. The notation *j* was not to be used. It is rather fortunate that we need only the first seven shells, and hence only seven

Shells Subshells Telescoped out

4–5 *Subshells and their names*

labels for subshells, to completely classify all known atoms in their ground states. Otherwise, we would be quickly coming around to another *p!*

I hope that I have convinced you of an almost total arbitrariness in choosing the system of names for atomic energy levels. There is nothing to understand or not understand about the spectroscopic notations. In any case, the standard labeling of subshells is (1s), (2s, 2p), 3s, 3p, 3d), (4s, 4p, 4d, 4f), (5s, 5p, 5d, 5f, 5g), and so on.

The schematic drawing of energy levels in FIG. 4–5 is not drawn to any particular scale, and energy differences are too ideally spaced. In reality there are wide variations in the distance between subshells. One of several realities is the one in which the gaps between two adjacent subshells become wider at higher ranked shells; that is, the energy gap between the 4s and 4p subshells is wider than that between the 3s and 3p subshells, or for that matter than that between the 3p and 3d subshells. As a result the 4s subshell has a lower energy level than that of the 3d subshell, as shown in FIG. 4–6. Such "crossover" patterns repeat themselves in higher levels.

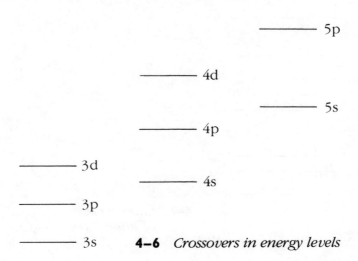

4–6 *Crossovers in energy levels*

As previousely mentioned, the quantum theory of atoms executes a set of well-defined computational steps of successive approximations and yields the energy levels for electrons at each of these subshells—results that are extremely accurate and in excellent agreement with experimental data. Several effects of the interaction among electrons, averaged over their distribution inside an atom, are added to the main backbone of one-electron atom analysis. Just to mention one example, take the magnetic influence one electron exerts on the other. As learned in chapter 3, a circular flow of electric current acts like a small magnet. Every electron, therefore, acts like a tiny magnet since it has a circular orbit. In a zirconium atom with 40 electrons zipping around in ten different subshells, there are 40 tiny magnets affecting each other!

When all is said and done, the quantum theory of atoms yields a unique hierarchy of the energy level schemes for subshells of all many-electron atoms. This is one of its central results and is shown in FIG. 4–7. There are many crossovers of levels and again it is only schematic, not drawn to any particular scale.

As you can see, even if subshells were ordered in an alphabetical sequence, it would have been all jumbled up anyway! It is an extremely boring and apparently meaningless picture: A bunch of

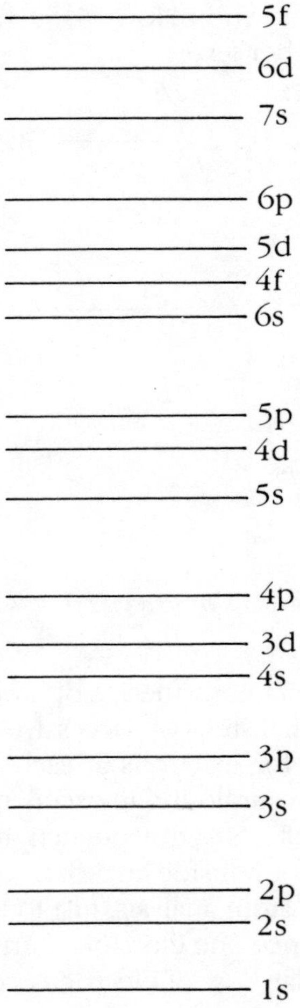

—————————— 5f

—————————— 6d

—————————— 7s

—————————— 6p

—————————— 5d
—————————— 4f
—————————— 6s

—————————— 5p
—————————— 4d

—————————— 5s

4–7 *Energy levels of subshells of atoms*

—————————— 4p

—————————— 3d
—————————— 4s

—————————— 3p

—————————— 3s

—————————— 2p
—————————— 2s

—————————— 1s

horizontal sticks laid out in a random vertical pattern, looking like an oversized Universal Product Code, and the sequence of labels that must have been cooked up by a malfunctioning decoding machine! Filling up the vertical pattern by electrons according to a set of strict rules, to be discussed in the next section, is to build up all known atoms in the universe. It is the basis for a "universal atomic code."

The housing rules for electrons

One atom differs from another by the number of electrons it consists of in its natural ground state of a stable configuration. The number of electrons matches exactly the number of protons inside its nucleus.

Atoms are, therefore, electrically neutral. Each atom is uniquely characterized by a number specifying the number of electrons it contains. This number is called the *atomic number* and denoted by a capital Z. The atomic number of a hydrogen atom is 1; it is 2 for a helium atom and 40 for a zirconium atom.

　　The quantum theory of atoms yields a set of very exacting rules that dictates not only the order of occupation, but also the number of electrons allowed per subshell. The total number of electrons and the way in which they fill subshells define, therefore, each atom and all of its properties. The rules of occupation, the housing rules, are hence of paramount importance to understanding the ways of atoms, molecules, biomolecules, solids, and the rest of all matter. These rules can be incorporated, for our purposes, in the following two rules:

　　Rule One　Electrons occupy subshells in such a manner as to result in the lowest possible energy for an atom. They start filling up from the bottom, the 1s subshell, and work their way up according to Rule Two.

　　Rule Two　Occupancy at each subshell is limited to a characteristic maximum number the subshell can accommodate. This maximum occupation number depends only on the ascending sequence of s, p, d, f, . . . , and not on the rank of a shell—that is, 1, 2, 3, 4,. . . . In other words, the subshells 1s, 2s, 3s, and so on are limited by the same number and so are 2p, 3p, 4p, and so on. As listed in TABLE 4–1, the pattern of the maximum occupation number corresponds to doubling of an ascending sequence of odd numbers—that is, 1, 3, 5, 7, . . . multiplied by two. This rule is absolutely exact.

　　In these two rules you now have the power and instruction to put together, one by one, all known atoms that make up the physical world around us. The first electron (hydrogen) goes into the lowest subshell, 1s, which gets filled up by two electrons (helium). The third electron must be placed into the 2s level (lithium), which gets filled up by the fourth electron, two in 1s and two in 2s (beryllium). The next six electrons, from the fifth through the tenth, occupy the 2p level until that level is filled up (boron, carbon, nitrogen, oxygen, flourine, and neon). Then you go on to the 3s and 3p levels.

　　Suppose you are standing in front of a system of 18 wall shelves. Each shelf is labeled by some nonsensical code made up

TABLE 4–1 *Maximum number of electrons allowed at each subshell*

Each subshell labelled	can accommodate only up to this many electrons
s	2
p	6
d	10
f	14
g	18
.	.
.	.

of numerals and four letters, s, p, d, and f. You are to place books on these shelves, starting from the bottom shelf, according to a rule, two books each on any shelf marked by s, six books each on any shelf marked by p, and so on. Now what can be simpler than that?

TABLE 4.2 lists the 18 subshells with the corresponding limit for maximum occupation at each level, and cumulative sums of the number of electrons completely filling up each level. The highest atomic number is 112, which is more than sufficient to accommodate atoms in their ground states up to 105. Atoms with atomic numbers of 2, 10, 18, 36, 54, and 86 are exceptionally noninteractive and are referred to as *inert gases*. Can you spot a pattern for them?

Counting up from the bottom in TABLE 4.2, we can now build atoms, one by one, just as easily as stacking up Lego® blocks. Let's begin with the first one, Z being equal to 1:

Z = 1 The atomic number is one, meaning one electron and one proton, the good old hydrogen atom. The lone electron occupies the 1s level, and the spectroscopic notation designating its electron configuration is $1s^1$. The superscript indicates the actual occupancy.

The notation $1s^1$, contains a lot more information than the name hydrogen. As long as you know its electron configuration, in fact, you don't really care what the atom is called. It is $1s^1$ in any language. Don't you think that the name *liquified 1s¹* sounds a lot more technical than a mere *liquid hydrogen?* Since the 1s level can

TABLE 4–2 *Orderly filling up of atomic levels*

Subshells	Maximum number of occupation	Cumulative sum, when fully filled
5f	14	112
6d	10	98
7s	2	88
6p	6	86 (Radon, Rn)
5d	10	80
4f	14	70
6s	2	56
5p	6	54 (Xenon, Xe)
4d	10	48
5s	2	38
4p	6	36 (Krypton, Kr)
3d	10	30
4s	2	20
3p	6	18 (Argon, Ar)
3s	2	12
2p	6	10 (Neon, Ne)
2s	2	4
1s	2	2 (Helium, He)

house up to two electrons, the $1s^1$ configuration has one unfilled vacancy, and this makes a hydrogen atom eager to accept an electron from a nearby atom, whether it is another hydrogen atom or a carbon atom, for example. To wit, the $1s^1$ is chemically active.

Z = 2 This atom has two electrons and two protons. The two electrons fill up the 1s level and happily hang out a sign saying that "Sorry, but no vacancy in 1s. We are $1s^2$." We have already met this fellow in FIG. 4–1, the helium atom. An atom with its outermost subshell completely filled is very stable. It has no need and hence no desire to pull electrons in from other atoms, and is not very active.

As indicated in TABLE 4–2, an inert gas atom has all its subshells completely filled, up to and just prior to the next-ranked *s* subshell. An argon atom has a completely filled 3p subshell, just below a 4s subshell, and likewise for other members of this group. Helium is the queen. It doesn't even form molecules with its own kind under normal conditions.

Z = 3   We have already met this one in FIG. 4–2. Its spectroscopic name is $1s^2\,2s^1$. The notation should be self-evident by now. The one unfilled vacancy in the 2s level makes this lithium atom behave in a manner very similar to a hydrogen atom, as was noted in the one-electron atom analogy.

Now the fun starts, the atomic fun. It can be very addictive! There is absolutely nothing to it. Anyone can play. All you need are the two rules and the hierarchy of ascending order of levels. We can write down the spectroscopic names up to, say, Z = 10 in half a breath.

Z =	4	beryllium	$1s^2\,2s^2$
	5	boron	$1s^2\,2s^2\,2p^1$
	6	carbon	$1s^2\,2s^2\,2p^2$
	7	nitrogen	$1s^2\,2s^2\,2p^3$
	8	oxygen	$1s^2\,2s^2\,2p^4$
	9	flourine	$1s^2\,2s^2\,2p^5$
	10	neon	$1s^2\,2s^2\,2p^6$

This *system is* not only a lot more systematic, but in fact a lot easier than license plates, nine-digit zip codes, Social Security numbers, or some of the IRS instructions to fill out a Form 1040.

In TABLE 4–3 are the electron configurations of atoms up to Z = 36. Certain irregularities begin to crop up as you go to higher atomic numbers. Chromium and copper have a slight irregularity in filling up the 4s and 3d levels, as indicated in the table. You can just write down the spectroscopic name for an atom simply by filling in the appropriate superscripts. A krypton atom is otherwise known as $1s^2\,2s^2\,2p^6\,3s^2\,3p^6\,4s^2\,3d^{10}\,4p^6$. No sweat!

As a general rule, most chemical properties of atoms are attributable to the degree of occupation, or the number of vacancies, of the s and p subshells. These vacancies determine to a large extent how atoms stick together to form molecules, compounds, DNAs, solids, and semiconductors.

One other aspect of atomic structures is worth discussing at this point. In chapter 2, we stated that the sizes of atoms are fairly constant from the smallest to the largest atom, varying very little in a narrow range between 1 and 3 angstroms, 1 angstrom being equal to 10^{-10} meters. Now, this property of atomic size does not seem to jibe all that well with what we have been discussing in this section. At higher atomic numbers, electrons are placed at higher subshells and this

TABLE 4–3 *Electron configurations of atoms up to Z = 36*

			1s	2s	2p	3s	3p	4s	3d	4p
1	H	Hydrogen	1							
2	He	Helium	2							
3	Li	Lithium	2	1						
4	Be	Beryllium	2	2						
5	B	Boron	2	2	1					
6	C	Carbon	2	2	2					
7	N	Nitrogen	2	2	3					
8	O	Oxygen	2	2	4					
9	F	Fluorine	2	2	5					
10	Ne	Neon	2	2	6					
11	Na	Sodium	2	2	6	1				
12	Mg	Magnesium	2	2	6	2				
13	Al	Aluminum	2	2	6	2	1			
14	Si	Silicon	2	2	6	2	2			
15	P	Phosphorus	2	2	6	2	3			
16	S	Sulfur	2	2	6	2	4			
17	Cl	Chlorine	2	2	6	2	5			
18	Ar	Argon	2	2	6	2	6			
19	K	Potassium	2	2	6	2	6	1		
20	Ca	Calcium	2	2	6	2	6	2		
21	Sc	Scandium	2	2	6	2	6	2	1	
22	Ti	Titanium	2	2	6	2	6	2	2	
23	V	Vanadium	2	2	6	2	6	2	3	
24	Cr	Chromium	2	2	6	2	6	1	5 (instead of 2 and 4)	
25	Mn	Manganese	2	2	6	2	6	2	5	
26	Fe	Iron	2	2	6	2	6	2	6	
27	Co	Cobalt	2	2	6	2	6	2	7	
28	Ni	Nickel	2	2	6	2	6	2	8	
29	Cu	Copper	2	2	6	2	6	1	10 (instead of 2 and 9)	
30	Zn	Zinc	2	2	6	2	6	2	10	
31	Ga	Gallium	2	2	6	2	6	2	10	1
32	Ge	Germanium	2	2	6	2	6	2	10	2
33	As	Arsenic	2	2	6	2	6	2	10	3
34	Se	Selenium	2	2	6	2	6	2	10	4
35	Br	Bromine	2	2	6	2	6	2	10	5
36	Kr	Krypton	2	2	6	2	6	2	10	6

implies a larger radius, for example, the sixth radius for an electron in the 6s level. It would appear to be reasonable to expect that atoms with a high Z could be 10, 20, and even 40 times as large as a hydrogen atom.

Here we witness one of the many quantum physical deftnesses of atoms. A higher atomic number also means many more protons in its nucleus, and the attractive force between a nucleus and an electron is that much stronger. What this does is to pull the whole pattern of radii closer toward the center of an atom, as illustrated in FIG. 4–8. Atom B with its outermost electron at the fourth radius, corresponding to levels 4s, 4p, 4d, or 4f, has about the same size as atom A, whose outermost radius is the second one, with its electron occupying either the 2s or 2p subshell.

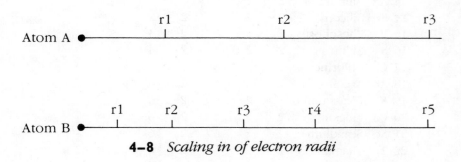

4–8 *Scaling in of electron radii*

5

✳

Molecules: societies of atoms

YOU HAVE JUST COME THROUGH, without a scratch, what is perhaps the most difficult chapter in this book. A conventional wisdom rules out such exposition of electron configurations from being included in a popular physics book. As you have just seen, it is not actually all that bad.

As you sail through the rest of this book, you will agree that it really is imperative to have such knowledge if you are to understand the technologies that depend heavily on the structure of atoms. Different arrangements of different numbers of electrons are what atoms are, and all their properties are dictated by electrons.

Molecular bonds

When two or more atoms are brought together within a close range, electrons in one of the atoms, especially those in the outermost subshell, begin to feel the pull of the nucleus of the other, and vice versa. Depending on the availability of vacancies in the outermost subshells of atoms involved, this mutual pull can lead to formation of molecules.

The situation is not unlike what might happen when two planetary systems, each with its own sun and a team of orbiting planets, are brought close together. The planet or planets of a system that are farthest out, such as Pluto of our Solar System, might just be pulled away and become a planet to the other sun, or the two systems might

5–1   *A planet around two suns*

coalesce into a single system comprised of two centers of force, with some planets going into orbits around two suns. Think what such an orbit would do to the change of seasons and the natural clocks of living things.

Something similar to a planet orbiting around two suns takes place with electrons when atoms coalesce and stick together to form various molecules, which is the subject we now come to discuss. A molecule can consist of only two atoms, such as a hydrogen or nitrogen molecule, or it can be built up from hundreds of thousands of atoms. No matter what the size, the only force holding them together is the electric force. It is the only agent by which atoms from molecules, giant molecules, liquids, solids, and all other matter, including apparently life itself. A *gene* itself, the portion of a DNA or a group of DNAs that contain genetic codes, is an arrangement of molecules held together by electric forces.

The way in which atoms form molecules is precisely controlled by the electron configurations of the constituent atoms. The electrons residing in inner subshells (closer to the nucleus) do not play any direct role, except to reduce the net effective positive charges of the central core, as with a lithium atom.

Almost all chemical properties are determined by the electrons, or their vacancies, in the outermost subshell and, to a lesser extent, in the second outermost subshell. Generally speaking, we can summarize the rules for forming molecules as follows:

Rule One   Subshells that are completely filled to their maximum capacity are fairly stable. By the same token partially filled subshells tend to be chemically active to reach that stability. Atoms with vacancies in their outermost subshell, and sometimes in the second outermost subshell, are very active in forming molecules.

Rule Two   Vacancies in any s- and p-subshells lead to very active molecule formation. Almost all *molecular bonds,* or glues holding atoms together, involve electrons in s- and p-subshells, of which 12 out of 18 are shown in FIG. 4–7.

Molecular bonds come in a variety of names—the ionic bond, the covalent bond, the hydrogen bond, and so on—all referring to slightly different arrangements of electrons by which atoms hook up with each other to form molecules. The basic mechanism for all types of bonds is the same as the electric attraction. We are not going to be overly concerned with precise definitions of these bonds, especially since in many instances the distinction between them gets blurred. Some atoms join up with others in such a way that more than one definition applies. Generally speaking, the ionic and covalent bonds are *intramolecular;* that is among the constituent atoms within a molecule they are forming. The hydrogen bond refers to an *intermolecular* bond; that is, between molecules.

A prime example of an ionic bond is the way in which a sodium atom and a chlorine atom form a molecule called sodium chloride, the chemical name for table salt. The whole situation becomes transparently obvious if you now look at the electron configurations of these atoms, which are listed in TABLE 5–1 with those for their immediate neighbors. We have already met neon and argon, with atomic numbers of 10 and 18 respectively. These two elements are members of a group of atoms called the inert gases, being very inactive and very stable.

TABLE 5–1 *Sodium, chlorine, and their neighbors*

Atomic number	Name	Symbol	Electron-configuration
10	Neon	Ne	$1s^2\ 2s^2\ 2p^6$
11	Sodium	Na	$1s^2\ 2s^2\ 2p^6\ 3s^1$
12	Magnesium	Mg	$1s^2\ 2s^2\ 2p^6\ 3s^2$
16	Sulphur	S	$1s^2\ 2s^2\ 2p^6\ 3s^2\ 3p^4$
17	Chlorine	Cl	$1s^2\ 2s^2\ 2p^6\ 3s^2\ 3p^5$
18	Argon	Ar	$1s^2\ 2s^2\ 2p^6\ 3s^2\ 3p^6$

With its outermost subshell, 3s, only half filled, a sodium atom can do one of two things to reach a relative stability: either gain an extra electron, thus filling up the 3s level to a full $3s^2$ configuration, or get rid of the $3s^1$ electron and fall back on the fully filled $2p^6$ configuration. The neutral sodium atom will turn into a negatively charged sodium ion with a magnesiumlike configuration in the first option, and into a positively charged sodium ion with a neonlike configuration in the second option. In both cases the nucleus remains the same with 11 units of positive charges. *Ion* refers to an electrically charged state of a neutral atom, which occurs when a neutral atom gains or loses one or more electrons under some conditions. Since a neonlike atom represents a very stable configuration, a sodium atom prefers to lose its lone electron, if it can find a receiver.

Losing an electron doesn't do much good, however, for a chlorine atom, $3p^5$ becoming $3p^4$ with two vacancies instead of one. A chlorine atom can, on the other hand, turn into a very stable argonlike configuration by accepting an electron, thereby filling up its 3p subshell to a full $3p^6$ and becoming a negatively charged chlorine ion. With a chlorine atom willing to receive and a sodium atom ready to donate an electron, you can guess the rest. The electron makes a jump from a sodium atom to a chlorine atom and, presto, the positive sodium ion and the negative chlorine ion attract each other, forming a sodium chloride molecule held together by an *ionic bond*.

Covalent bonds: sharing of electrons

The most common way for atoms to form molecules is through a mechanism referred to as a covalent bond. Two or more atoms with vacancies in their outermost subshells mutually share one or more pairs of electrons among themselves, filling up their vacancies part of the time—a sort of do-the-best-you-can partial compromise.

The covalent bond between a hydrogen, $1s^1$, and a lithium, $1s^2 2s^1$, atom is schematically shown in FIG. 5–2. The pair of electrons encircled by a dotted line represents the sharing. The electron from a hydrogen atom spends part of its time "filling" up the 2s level of a

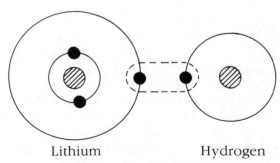

Lithium   Hydrogen

5–2 *The covalent bond between a hydrogen and a lithium*

lithium atom, keeping it happy, and likewise the electron in 2s level partially fills up the 1s level of the hydrogen atom. Both electrons belong to both atoms, and in this process of being shared, they spend more time located in between and thereby forming a bridge of attractive forces between the two atoms. This bridge is the *covalent bond.* A molecule made up of a hydrogen and a lithium atom, bonded by one pair of shared electrons, is denoted by inserting a short horizontal line between the atomic symbols: H–Li.

Covalent bonds account for almost all organic molecules as well as some very relevant inorganic molecules that make up materials for semiconductors and superconductors. It is easy to see that this simple sharing of electrons among the family of atoms can lead to many varieties of molecules. Specifically we are dealing with vacancies that number from one to five, the maximum number of vacancies in any p-subshell being five. An atom with two vacancies can form a molecule either with another atom with two vacancies or with two atoms of one vacancy each. An atom with three vacancies has more options: with another atom having three vacancies, with an atom of two and another of one vacancy, or with three atoms of one vacancy each. Let us look at some familiar and simple examples of molecules formed by covalent bonds among the four elements that make up 98% of our body mass: hydrogen, oxygen, nitrogen, and carbon. Their relevant properties are listed in TABLE 5–2. The number of covalent bonds corresponds to the number of vacancies in the outermost subshell of an atom.

TABLE 5–2 *Four principal elements of our body*

Atomic number	Name	Electron-configuration	Number of covalent bonds	% by mass in our body
1	Hydrogen	$1s^1$	1	10%
6	Carbon	$1s^2\,2s^2\,2p^2$	4	19%
7	Nitrogen	$1s^2\,2s^2\,2p^3$	3	7%
8	Oxygen	$1s^2\,2s^2\,2p^4$	2	65%

A hydrogen atom with one covalent bond A hydrogen atom is the simplest example, involving one vacancy in an s-subshell, as we have already seen in FIG. 5–2. A hydrogen molecule, denoted H_2, is formed by a covalent bond between two hydrogen atoms, each of which has the $1s^1$ configuration. Each "borrows" the electron of the other so that each can claim the full $1s^2$ status, at least partially. A picture of a hydrogen molecule is like the one in FIG. 5–2, except that you mentally remove the inner shell with two electrons from the atom on the left.

After hydrogen and lithium, $1s^1$ and $1s^2 2s^1$, the third atom with one covalent bond is sodium, $1s^2 2s^2 2p^6 3s^1$. The sodium also takes part in an ionic bonding, as we have already seen. Molecules formed out of these atoms with one covalent bond are denoted by H–H, Li–Li, Na–Na, Li–H, Na–H, and so on.

An oxygen atom with two covalent bonds With its outermost subshell, 2p, filled only by four electrons out of the maximum six, an oxygen atom has two vacancies. By sharing two of its four outer electrons, it can form a molecule with another atom of two vacancies or with two atoms of one vacancy each. The simplest example of the former case is an oxygen molecule, 0_2, and so the latter case is the good old H_2O.

Oxygen molecule	0_2	$0 = 0$
Water molecule	H_2O	H-O-H

As mentioned in chapter 3, the electric charge distribution in a water molecule is slightly polarized, like an electric dipole. Because of its nucleus of eight units of positive charges, the oxygen exerts a strong pull and the electrons being shared between the oxygen and hydrogen atoms end up spending more time hanging around near the oxygen nucleus. This turns the water molecule into a configuration with two slightly positively charged ends sticking out. This is the basis for the hydrogen bond, which we will describe in the next section.

A nitrogen atom with three covalent bonds With the configuration of $2p^3$, a nitrogen atom can exercise three-bondsmanship. Two familiar examples are a nitrogen molecule and an ammonia molecule.

Nitrogen molecule	N_2	$N\equiv N$
		$\diagup$ H
Ammonia	NH_3	H — N
		$\diagdown$ H

An ammonia molecule is depicted in FIG. 5–3. The 2p subshell of the nitrogen atom gets "filled" with one electron each from three hy-

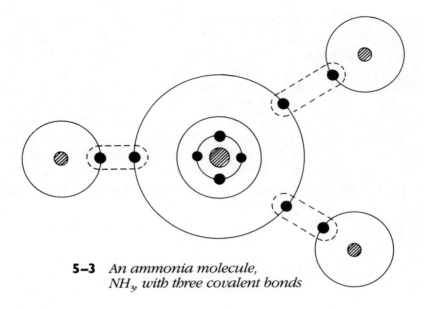

5–3 *An ammonia molecule,*
NH_3, with three covalent bonds

drogen atoms, and each hydrogen atom feels like a respectable $1s^2$. Very intelligent, sociable, and harmonious.

An ammonia molecule with one hydrogen atom missing is called an *amine group,* denoted by —NH_2, the "missing atom formation." It has one covalent bond open and ready to link up with other atoms. When linked up with a carbon atom it becomes one of the principal constituents in all 20 different amino acids, the substances that form all different varieties of proteins. It is no wonder that we are so full of nitrogen, carbon, hydrogen, and oxygen!

A carbon atom with four covalent bonds Now we come to the all important carbon, its outermost subshell $2p^2$ showing four vacancies. It needs four electrons to participate in sharing and a carbon atom recruits the two electrons in the second outermost subshell, $2s^2$, to join up with the two electrons in the 2p level in forming four covalent bonds. This is a slight exception to the rule, but the pattern repeats itself with an atom whose two outer subshells are $3s^2$ and $3p^2$, all four electrons forming covalent bonds. This copy of a carbon atom is none other than a silicon atom. We will discuss this point further in the next chapter dealing with semiconductors.

Carbon atoms serve to differentiate organic from inorganic molecules and provide a central basis for all molecules of life. Some simple and well-known examples of molecules containing carbon atoms are shown in FIG. 5–4, in which each line between the atomic symbols represent one covalent bond, or a pair of shared electrons.

The central core of a benzene molecule, the hexagonal arrangement of covalent bonds called a *benzene ring* with some carbon atoms replaced by nitrogen atoms, is a central foundation for many organic molecules, including those that form parts of molecules of life. You see the hexagonal logos and trademarks in so many places throughout the chemical, pharmaceutical, and biotech industries. These hexagonal rings anchor a multitude of complex molecules, which include four nitrogen compounds, called *bases*. The four

Methane CH_4

Ethane C_2H_6

Carbon dioxide CO_2

Benzene C_6H_6

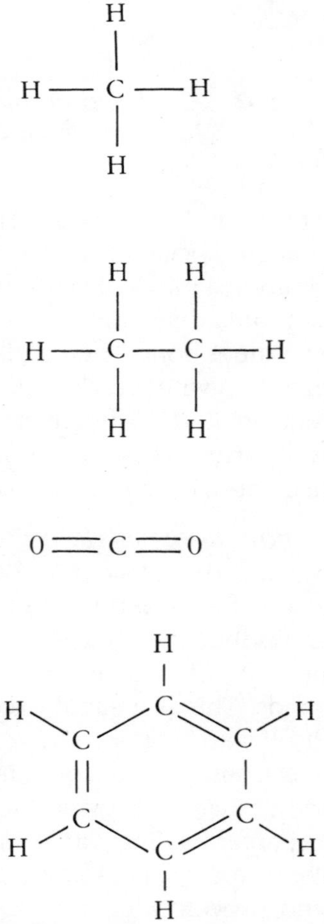

5–4 *Carbon and its bonds*

bases—adenine, cytosine, guanine, and thymine—shown in FIG. 5–5 are the basic ingredients of a DNA molecule, the basic molecule of life. A DNA molecule is made up of thousands of these bases in many different permutations. The four bases, usually denoted simply as A, C, G, and T, and the covalent bonds holding them together are tabulated in TABLE 5–3.

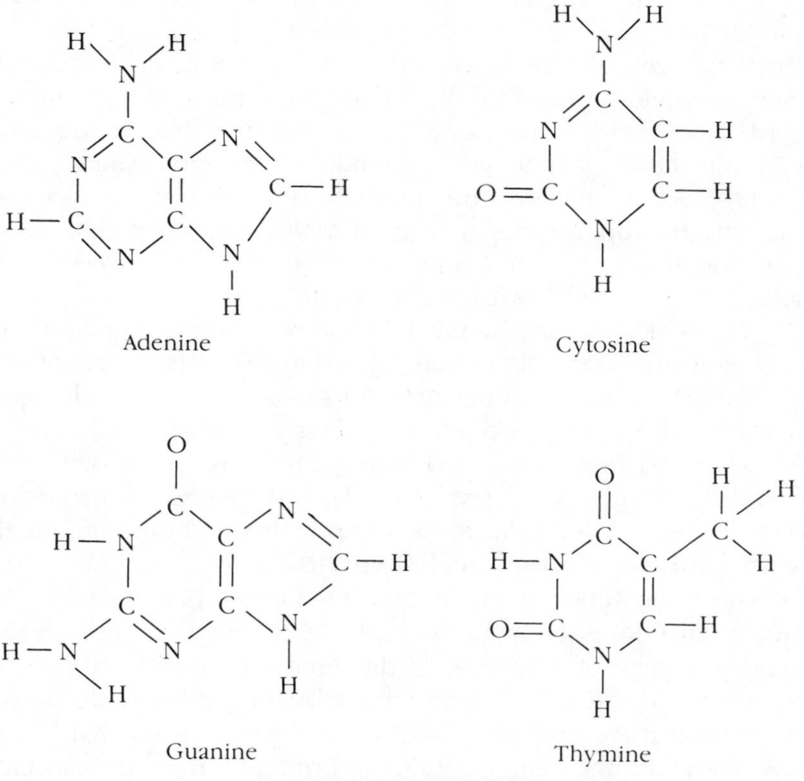

Adenine

Cytosine

Guanine

Thymine

5–5 *The four bases of DNA*

TABLE 5–3 *The four bases and their contents*

Name	Symbol	Total number of covalent bonds	Total number of atoms	H	O	N	C
Cytosine	C	16	13	5	1	3	4
Guanine	G	21	16	5	1	5	5
Adenine	A	20	15	5	0	5	5
Thymine	T	18	15	6	2	2	5

Hydrogen bonds of water and life

We hear often about the hydrogen bond. It plays a singularly important role as the mechanism by which the two strands of a double helix of a DNA molecule are held together. It is also what holds water molecules together in an ice crystal. It is not a new kind of a mechanism, but rather a special name for a special attraction between two or more molecules.

As we have already discussed in the case of a water molecule, many molecules are electrically *polar;* that is, because of an uneven distribution of electrons among the constituent atoms, some regions of a molecule are slightly positive while other regions slightly negative. Oppositely charged ends of such molecules will attract each other and link up, forming a chain of molecules. The mechanism is in fact identical to the ionic bond, except that the former is between molecules and is much weaker than the latter.

A class of such binding in which the positively charged end of a polar molecule happens to be an ionized hydrogen atom, a *proton,* is referred to by a special name: the *hydrogen bond.* As we explained in the last section, each pair of electrons being shared between an oxygen atom and one of the two hydrogen atoms spend more time around the oxygen atom, giving a polar distribution of charges. Ice crystal is formed in a solid form by the hydrogen bonds among the electrically polar water molecules. See FIG. 5–6.

The prominence of a hydrogen bond is due perhaps to the fact that it is what holds the two strands of a DNA molecule together. Each strand is a chain of thousands of different permutations of the four bases, A, C, G, and T. Coded into this sequence of bases are instructions to construct a matching sequence of amino acids; that is, proteins. These are the genes, holding information of the development of a life. Molecular structures of these four bases are such that base A can only be bonded with base T with the help of two hydrogen bonds, while base C can only couple with base G by using three hydrogen bonds. This rule appears to be very precise. A two-prong pairing between A and T and a three-prong pairing between C and G is shown in FIG. 5–7, where the hydrogen bonds are indicated by dashed lines.

When the sequence of bases on one of the strands of a DNA is determined, so is the matching sequence of bases on the other strand, as a result of this specific pairings between the bases, as shown schematically in FIG. 5–8. Under the right conditions, the relatively

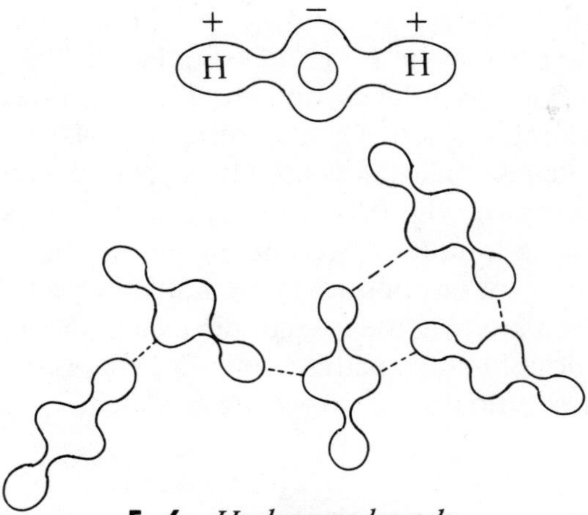

5–6 *Hydrogen bonds of water molecules*

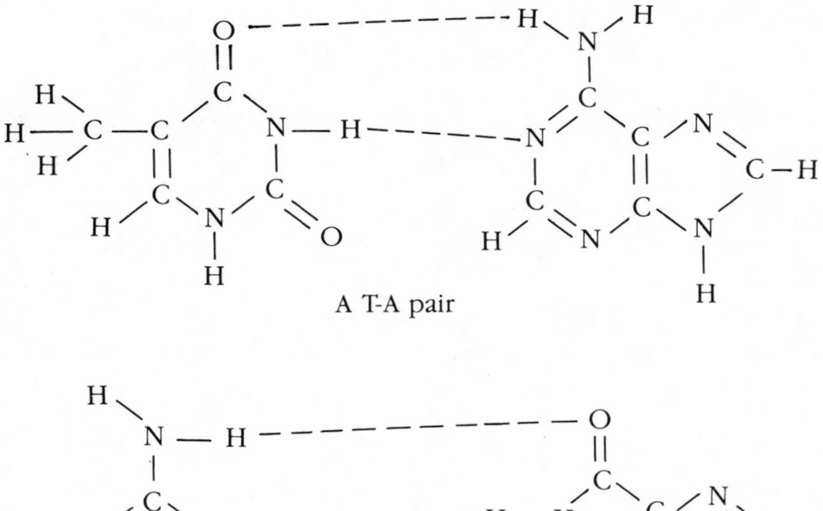

A T-A pair

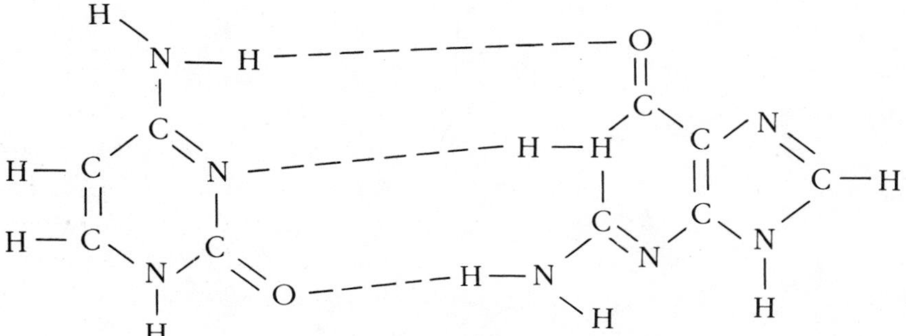

A C-G pair

5–7 *The two- and three-prong hydrogen bonds for thymine-adenine and cytosine-guanine pairs* weak hydrogen bonds can be broken and the two strands begin to separate, starting from one end and completely "unzipping" themselves. Each strand, completely unwound and split apart, begins to pick up a matching sequence of the complementary bases within a cell material and, presto, we now have two DNA molecules, each containing one of the two original strands. The process repeats again and again: two become four, four become eight—the process of replication! The mechanism holding together water molecules is identical to the mechanism keeping the two strands of a DNA molecule twisted in a form of a double helix. Is this something that is truly amazing or something that should not be surprising at all?

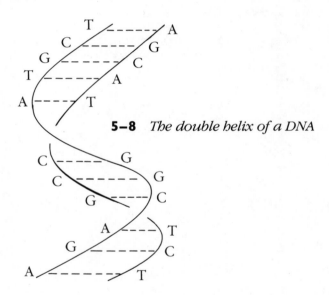

5–8   *The double helix of a DNA*

6

✳

Semiconductors, chips, and all that

OF THREE STATES OF MATTER, solid, liquid, and gas, solids are the most tangible things that define the physical world in which we live and work. To us, the land inhabitants, such words as *solids, things,* and *objects* are almost synonymous. Intuitively we expect a thing to have a definite shape and size, a hardness you can feel, and occupy a well-defined location. Our natural instinct to analyze the physical world around us in terms of point particles is deeply rooted in our concept of solids. We have done rather well with solids over the years—the Stone Age, the Bronze Age, the age of steel, and now the age of semiconductors and chips.

The super, regular, and semi of conductors

Solids are densely packed chunks of atoms, molecules, and other compounds, some made of a single element such as a piece of copper or diamond but most a mixture of various atoms. Atoms and molecules are packed so closely that an average separation between neighbors is roughly the same as their sizes. At this close range the distinction between intramolecular and intermolecular bonds gets very blurred. An analysis in terms of individual atoms gives way to a collective and statistical approach, but it is through the knowledge of the electron configurations of atoms that we have come to understand solids and have launched the high-tech of today.

One of the two principal characteristics of a solid is its *lattice* structure, which refers to the patterns of arrangement of atoms inside it. Consider a wall covered by wallpaper, displaying an unending repetition of some precise geometrical pattern, and imagine it in three dimensions. Even though we do not yet have a clear understanding of the recently discovered so-called high-temperature ceramic superconductors, it is generally believed that the lattice structure of the copper oxides, the main ingredients of these ceramics, is largely responsible for the observed superconductivity.

The other major characteristic of solids is the number and distribution of free electrons. Strictly speaking, electrons are *free* only when they are totally unbound from any atomic nucleus and exist all by themselves, as mentioned in chapter 2. Electrons in solids are bound so weakly to a bunch of atoms in a lattice that they are referred to as free electrons. Because the close packing of atoms mutually cancels out the pushes and pulls, these electrons, mostly coming from the outermost but sometimes the second outermost subshells, are quite free to move about even when only a slight nudge is applied, such as a connection to a battery. The two principal characteristics, the lattice structure and the free electron density, determine all physical properties of solids, which include among others the electrical conductivity.

An electric current is a flow of electric charges, just as a river current is a flow of water molecules. By a convention adopted long ago, the direction of the current flow is taken to be the direction in which positive charges would flow, from positive to negative terminals. The direction of the actual flow of electrons through the interior of a wire is, because of their negative charges, opposite to the direction adopted by this old convention. It is not a big deal, but something to keep in mind when we discuss the positive- and negative-type semiconductors later.

Whether a material is a good conductor of electricity or not is determined by the relative density of free electrons it contains. The total number of free electrons depends on the density of atoms per unit volume of a material and the net number of free electrons you can expect from each atom. With the exception of a small number of elements called semiconductors, most solids belong to one extreme or another; they are either very good conductors or very good insulators.

All good conductors are metals and the best among them are silver, copper, and aluminum. Of the three, silver is the best conductor. House wirings are not done, however, in silver for an obvious reason. From TABLE 4–3 we can read off the electron configurations of an aluminum atom, Z = 13, and of a copper atom, Z = 29, to be $1s^2$ $2s^2$ $2p^6$ $3s^2$ $3p^1$ and $1s^2$ $2s^2$ $2p^6$ $3s^2$ $3p^6$ $3d^{10}$ $4s^1$, respectively. A slight irregularity in a copper atom results in having the 3d level filled up first, leaving out one electron in the 4s level. The last two subshells in a silver atom, Z = 47, not shown in TABLE 4–3, are $4d^{10}$ $5s^1$, repeating the same type of irregularity as in a copper atom.

In any case, these three conductors have the highest density of free electrons, one free electron per atom on the average. The "lonesome end" electron of each atom, $3p^1$, $4s^1$, and $5s^1$ respectively, contributes to the free electron pool. The availability of one free electron per atom, on the average, results in about 10^{22} free electrons per cubic centimeter (10 billion trillion of them). Since 1 cubic inch is equal to about 16.4 cubic centimeters, there are about 164 billion trillion free electrons contained in a cube of aluminum 1 inch by 1 inch by 1 inch.

Other good conductors include iron, nickel, and alloys such as brass and steel.

In good insulators, on the other hand—such as glass, quartz, paper, hard rubber, mica, and porcelain—almost all electrons are engaged in forming tight bonds. As a result, these atoms provide very little free electrons, and hence do not carry any useful amount of electric current. The density of free electrons for good insulators is typically about 100 per cubic centimeter, down by a factor of 10^{20} from a good conductor. This is tantamount to having no free electrons.

There is one group of elements, sometimes called the *diamond group,* that exhibits electric conductivity under certain circumstances. The group consists of carbon (diamond), silicon, germanium, and tin, all of which have the same bonding pattern: the four covalent bonds as discussed in the previous chapter. Pure solids of these elements contain free electrons with a typical density of about 10^{12}; that is, about 1 trillion per cubic centimeter, down by a factor of 10^{10} from that for a good conductor. These are the naturally occurring so-called *pure,* or *intrinsic,* semiconductors.

There is nothing really *semi-* about these solids, with the density of free electrons being approximately equal to one-tenth of one-billionth that for a good conductor. If anything, these intrinsic or pure semiconductors are closer to insulators. These naturally occurring semiconductors are, however, not the usual material generally referred to as semiconductors by everybody, as in *semiconductor industry* or *semiconductor trade agreements*. The wonder material that ushered in today's high-tech and the age of information is artificial. These so-called *doped* or *impure* semiconductors are manufactured by a controlled mixing another element into a base of a pure semiconductor.

Pure semiconductors

Being down by a factor of 10^{10} from a good conductor, the free electron density of an intrinsic semiconductor corresponds to the rate of about 1 free electron per every 10 billion atoms—a free electron by a rare accident so to speak, which cannot sustain any meaningful and practical amount of electrical current even by a standard of microelectronics. The pure semiconductors, however, provide basic frameworks of atomic latticework into which a controlled amount of other atoms, impurities, can be introduced. In order to understand the process of how impurities improve electrical conductivity, let us first look at the lattice structure of pure semiconductors in terms of the covalent molecular bonds discussed in the last chapter.

Almost all pure semiconductors are solids formed by atoms having the same pattern of forming four covalent bonds as carbon atoms. The four principal members of this group, called the *diamond* or *carbon group,* are: carbon, silicon, germanium, and tin. Each of these elements forms four covalent bonds in the manner discussed in the last chapter. A carbon atom is characterized by $1s^2\, 2s^2\, 2p^2$. Each of the four electrons, two in 2s and two in 2p, enters into a covalent bond with another electron from other atoms, "filling" up the four vacancies in the 2p level.

Following the standard notation in chemistry, we have indicated one such bond by a short line between the atomic symbols, such as shown in the examples in FIG. 5–4. In discussing the lattice structure of semiconductors, it is more convenient to explicitly show electrons by dots. For example, the diagram for a methane molecule, shown in FIG. 5–4, will be presented as in FIG. 6–1, in which only those electrons forming a bond are indicated.

In the example of a methane molecule, all eight electrons belong to all parties involved. All eight "fill" up the 2s and 2p levels of a

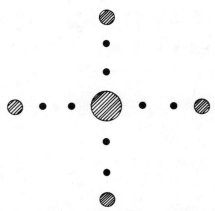

6–1   *A methane molecule revisited*

carbon atom part of the time. At the same time all four hydrogen atoms get "filled" up, two apiece.

The sequence of atoms that repeats this pattern of four electrons forming four bonds consists of those atoms whose two outer sub-shells are $2s^2 2p^2$, $3s^2 3p^2$, $4s^2 4p^2$, $5s^2 5p^2$, etc. The next one up from carbon is none other than silicon, atomic number 14 and list of $1s^2 2s^2 2p^6 3s^2 3p^2$, followed by germanium and tin. Once you know the simple system of the electron configuration of atoms, so many things about atoms and molecules become very clear and as easy as building things up with Lego blocks—these Lego blocks also have a simple system: blocks with two prongs, six prongs, and so on.

The lattice structure of a solid of pure carbon or pure silicon can be conveniently represented as shown in FIG. 6–2, for each set of four dots signifying the four electrons, either $2s^2$ and $2p^2$ a carbon atom or $3s^2$ and $3p^2$ for a silicon atom. Lest you get oversold on such a perfect and simple picture, keep in mind that such a picture is an abstraction. In the first place, everything is in three dimensions and some of the atoms can be lifted off the page, while still some others may be deep behind the page. There will be many irregular spacings, some bent and others twisted.

The picture shown in FIG. 6–2 is too much of a simplification. Nevertheless, it serves an important purpose of helping us visualize what goes on in the incredibly small microcosmos of atoms. So we just keep going merrily along, drawing little circles and putting little dots in between. It is, in fact, a little difficult to see how we get any

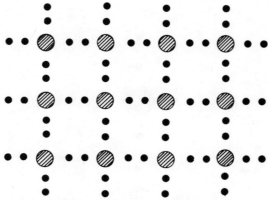

6–2   *A schematic representation of the lattice structure of a pure semiconductor*

free electrons at all out of such a lattice work, but irregularities allow free electrons at the rate of 1 free electron per 10 billion atoms, as mentioned previously.

One interesting observation can be made in passing. Of the four principal elements that make up the bulk of our body, carbon is the most characteristic ingredient of life. We are "carbons" after all. The atom closest to carbon in its structure is the silicon atom. Silicon forms silica, sand, quartz, rock crystals, and slicate rocks, and is one of the most abundant materials on Earth. After all these years, the "carbons" have developed high-tech and in so doing have discovered the usefulness of their closest atomic kin, the "silicons".

Doped semiconductors

Now we come to the artificial semiconductor, not wholly man-made but a manufactured product in which a controlled amount of carefully selected impurities is mixed into a pure semiconductor. In most cases of material processing, impurities are something to avoid or get rid of, but then there are those cases such as in steel and rubbers for tires where certain impurities improve the strength and quality. In the case of semiconductors, they improve the capability of carrying current. Although not as impossible as insulators, naturally occurring semiconductors such as diamond and silicon composites might just as well be insulators as far as meaningful practical applications are concerned. If the density of free electrons can be increased by injecting impurities of atoms having just the right kind of electron configuration, then it would be possible to come up with a material whose electrical conductivity lies somewhere between a good conductor and a pure semiconductor.

Now you might ask, why bother? What is the purpose of having such a partial conductor? Why go through all the trouble since we already have excellent conductors such as silver, copper, and aluminum anyway? In the technology of electronics many more devices than just good conductors *are needed.* In fact, all sorts of components with special functions are needed: things that cut down voltages, called *resistors,* components that can temporarily store electric charges and then discharge them gradually, called *capacitors,* amplifiers, and electronic switches. More importantly, we need materials that can pass currents under certain conditions and block currents under other circumstances; that is, "semi"conductors in the true sense of the word. Good conductors do only one thing: pass currents. Because it is somewhere between a good conductor and a pure semiconductor, the doped semiconductor is exactly such a material.

Many crucial electronic components can be made out of this new material not only more efficiently, but also much cheaper and, more importantly, much, much smaller. This is how the vacuum tubes went the way of the dinosaurs and microelectronics was born. All of this wouldn't have been possible without the knowledge of the atomic structures afforded us by the quantum physics of atoms.

Of the four elements—carbon, silicon, germanium, and tin— that make up the pure semiconductors, silicon and germanium are the most widely used base material to which impurities are embedded. The reasons are that they are sturdy, readily and inexpensively available, and convenient to handle. Both silicon and germanium have lattice structures identical to the one shown in FIG. 6–2, the four electrons coming from $3s^2\ 3p^2$ for silicon and $4s^2\ 4p^2$ for germanium. The atomic number of germanium is 32 and its full list of electrons is shown in TABLE 4–3; that is, $1s^2\ 2s^2\ 2p^6\ 3p^6\ 3s^2\ 3p^6\ 3d^{10}\ 4s^2\ 4p^2$. A slight irregularity contributes to filling up the 3d level before the 4s level for atoms from copper to krypton.

Now the trick is to look for an atom that has such an electron configuration so that, when it replaces one of the silicon or germanium atoms in its lattice, four of its electrons fit precisely into four slots and one electron is still *left over.* Well, you don't have to be a genius to realize that the answer is obvious; it is no trick at all to find such an atom. It is the one with either $3s^2\ 3p^3$ or $4s^2\ 4p^3$. From TABLE 4–3, you can see that it is either phosphorus, with atomic number 15, or arsenic, with atomic number 33. In either case, the third electron in the outermost p-subshell is the odd man out! It has no place to fit into and, presto, no choice but to become one free electron.

For many practical advantages of manufacturing technology, the pair most used consists of a silicon base and arsenic impurities. For every arsenic atom that mixes in and replaces a silicon atom, there is one free electron. This situation of an arsenic impurity is shown in FIG. 6–3, in which the arsenic and silicon atoms are designated by their notations, Ar and Si, and the odd-man-out electron is so indicated.

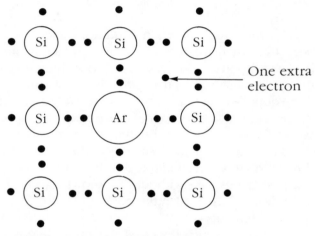

6–3 *An arsenic atom impurity*

Doping, as the process of introducing impurities is called, is done by diffusing an arsenic vapor onto a silicon wafer and baking it inside a high-temperature oven. We are talking about embedding impurities at an atomic scale and, needless to say, every step of manufacturing must be executed in a clean environment that is as close to a vacuum as possible—what is referred to in the industry as a *clean room*. Impurities are mixed in at an average rate of 1 arsenic atom in 1 million to 10 million silicon atoms. This rate translates into a free electron density of about 10^{15} to 10^{16} per cubic centimeter, which is enough to pass a small amount of current under the right conditions.

This arsenic-doped silicon is a typical example of what is called the *n-type semiconductor,* one of two types of impure semiconductors, the other being the *p-type semiconductor.* The n-type (*n* for negative) semiconductor is so called because the carriers of electric current are the free electrons donated by the impurity atoms and electrons carry negative charges. Now at first thought, the descriptive name, *n-type,* appears to be totally superfluous and redundant. We have never bothered with calling coppers and alluminums "n-type" conductors. What other types of conductors are there? They conduct current by virtue of the motion of electrons inside them, which are negatively charged. Why all of a sudden do we have to call these semiconductors n-type?

Although there can be only one type of good conductor, n-type, doped semiconductors can be obtained by another, complementary kind of doping that provides just as good a mechanism. It comes about this way. If doping can be done using atoms having one more electron than silicon or germanium, then it also can be done using atoms having one *less* electron then silicon or germanium. This second type of doped semiconductor is called the *p-type* (*p* for positive).

In the same way as for the n-type, let us look for atoms having one electron less than $3s^2\ 3p^2$ of silicon or $4s^2\ 4p^2$ of germanium. We don't have to look very far because the atoms that we want are the ones with $3s^2\ 3p^1$ and $4s^2\ 4p^1$. It is that simple. We cannot use the one with $3s^2\ 3p^1$ because it is none other than aluminum itself, a good old good conductor! We are left with the one with $4s^2\ 4p^1$, a gallium atom with atomic number 31. When a gallium atom replaces a silicon atom, the three electrons, $4s^2$ and $4p^1$, fit into three of the four slots and in contrast to the case of an arsenic atom giving off one free electron, the gallium atom leaves one slot vacant and unfilled as shown in FIG. 6–4. Many manufacturing processes prefer gallium impurities in a silicon base.

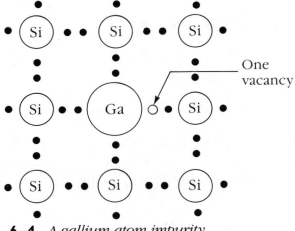

6–4 *A gallium atom impurity*

Instead of getting free electrons, with the gallium impurities we get unfilled vacancies in the sea of silicon atoms. These unfilled vacancies are called *holes*. When such a material is connected to a battery, a whole new game erupts among the electrons in the lattice. There are no movements of free electrons since there aren't any extra free electrons, but one of the electrons in the immediate neighborhood of a hole, realizing that there is a place to go, will jump over and fill that hole.

A second electron will jump over and fill the hole left by the first electron. Other electrons like what they see and before you know it every electron gets in on the new game in town, the electronic follow-the-leader. We have a movement of electrons, an electric current, just as good a current as the one in the case of n-type semiconductors. This is with just one hole. There are as many holes as gallium impurities. This is how a p-type semiconductor works, but where is the positive charge?

Consider the sequence shown in FIG. 6–5 indicating an orderly game of follow-the-leader by electrons, filling up holes in a highly organized manner. As each electron moves to the left to fill a hole in succession, what we have is a motion of a hole to the right! Place this page containing FIG. 6–5 about 4 feet away from you, and as you view the lines from top to bottom, you will see a motion of a hole to the right instead of electrons jumping to the left. In terms of the electric charge, a hole represents a double negative, a lack of negative charge, which appears to be a "positive" charge.

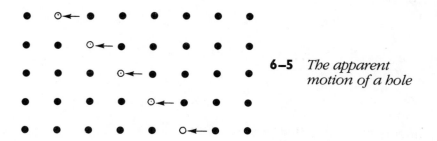

6–5 *The apparent motion of a hole*

A vapor of gallium atoms is diffused into a silicon base to attain about the same degree of impurity as with arsenic atoms—1 gallium atom embedded in about 1 million to 10 million silicon atoms. This contributes about 10^{15} to 10^{16} holes, or "positive" charges, per cubic centimeter. This gallium-doped silicon is the principal example of the p-type semiconductors.

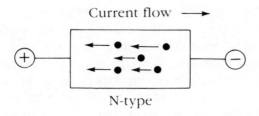

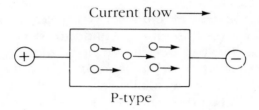

6–6 *The n- and p-type semiconductors*

Both n-type and p-type semiconductors allow flow of electric currents under certain conditions, in the same direction but with different mechanisms—one with electrons and the other with their vacancies. As shown in FIG. 6–6, in a p-type semiconductor the direction of the current flow is the same as that of the flow of holes, but opposite to that of the flow of electrons in a n-type semiconductor. The knowledge of the lattice structures of some solids has enabled us to create not one but two types of new materials by playing the atomic game of mixing in just the right kind of impurities. These new materials were to completely revolutionize the electronics industry.

Transistors and chips

Almost all basic components for electronic circuitry can now be built up by suitably combining n- and p-type semiconductors, the most important of which include such devices as diodes, transistors, and integrated circuits, otherwise known simply as *chips*. Let us discuss some salient features of these wonder components

One-part devices   First, let us consider the doped semiconductors by themselves. Although with much less ease than good metallic conductors, these materials do allow electric currents to pass, thus making them rather useful resistors by themselves. A good resistor is a poor conductor and vice versa, and the degree of resistance can be controlled by varying the amount of either gallium or arsenic to be mixed in. Now resistors are readily and cheaply available compared to other materials without the trouble of atomic impurities. Semiconductors as resistors come in handy since they enable an entire electronic circuitry to be fabricated on a single piece of silicon no larger than a baby's fingertip: the integrated circuits.

Two-part devices or diodes   As a typical two-part device, let us consider a very useful semiconductor device called a p-n junction diode. Two-part devices are called *diodes,* meaning two paths, and three-part devices *triodes,* meaning three paths. Of course, no one has ever heard of such a silly name as mono-odes!

A p-n junction diode is shown in FIG. 6–7 in such a way that it looks like a piece of n-type and a piece of p-type are pressed together in the middle. In practice, such a diode can be fabricated out of a single silicon base in which doping by arsenic and gallium is done at different regions, predominantly arsenic to the left and predominantly gallium to the right in the diagram.

In the first part of the figure, the p-type side is connected to the positive terminal and the n-type side to the negative terminal of a battery. The diode is thus *forward-biased*—just one of those fancy names that physicists love to toss out to the uninitiated. While connected this way, the free electrons from the n-type side move toward the positive terminal, to the right in the diagram; at the same time the holes from the p-type side migrate to the left, toward the negative terminal. FIGURE 6–7 shows that this situation combines the best of both types—both n- and p-type add their efforts and there is a nice flow of electric current from the positive to the negative terminals, from right to left in the figure. The diode semiconducts nicely, so to speak.

Now reverse the situation and have the p-type side connected to the negative terminal and the n-type side connected to the positive terminal. The diode is thus *reverse-biased.* In this method of connection is the worst of both worlds. As shown in the figure, the free electrons accumulate near the positive terminal, and the holes congregate at the other end. There is no flow of electric current. An extremely simple device, the diode will pass currents only in one direction: from p-type to n-type. This property makes a diode

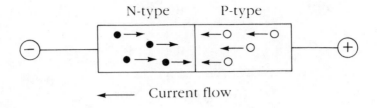

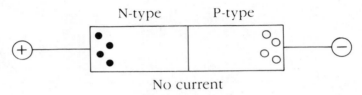

6-7 *A p-n diode*

a good *rectifier,* filtering out the alternating current into a direct current. The principle under which a diode works finds its greatest contribution, however, in a three-part device called a *transistor,* which is a sort of two diodes put together head to head.

Three-part devices, transistors This is the one, the one that was invented in 1948, the one that brings to the pinnacle the usefulness of semiconductors, and the one that was to profoundly change forever the way we handle and communicate information of our civilizations. The three-part device, called the *transistor,* a name shortened from "transfer resistor", first found applications rather modestly inside simple pocket calculators and hand-held transistor radios.

A transistor consists of a thin layer of one type of semiconductor sandwiched between the other types, either a p-n-p or a n-p-n arrangement as shown in FIG. 6-8A and B. The thin sandwiched part, whether it is a p- or n-type, is called the *base*, B, and the ends are called the *emitter* and *collector*, E and C. There are actually several things going on simultaneously inside a transistor when the three ends—E, C, and B—are connected to external sources of voltage; there are flows, and counterflows of electrons and holes.

As far as microelectronics is concerned, the importance of transistors lies in their switching capabilities. They are very small,

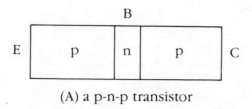

(A) a p-n-p transistor

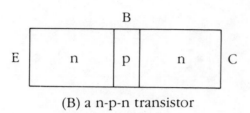

(B) a n-p-n transistor

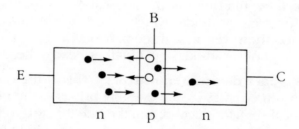

(C) Electrons move from E to C

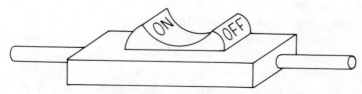

(D) The base, B, as a switch

6-8   *A transistor as a switch*

very fast, and very reliable electronic switching systems, forming the building blocks of all integrated circuits.

Let us see how this switching function comes about. We will use the n-p-n arrangement as shown in c of FIG. 6–8. The whole idea is to control—turn on and off—the flow of electrons from the emitter to the collector by an electrical signal supplied to the base. The base acts like a sort of a valve. E and C are connected to a voltage supply in such a way that the voltage at C is always higher than the voltage at E. If an independent electric signal is sent to B in such a way that the voltage at B is also higher than the voltage at E, the "diode" part, E and B, is forward-biased, and electrons move from E to B while the holes migrate from B to E. The base is made to be very thin so that the electrons coming into B from E can continue their motion, across the barrier between B and C and all the way toward C. We have a flow of current; the switch is on.

If, on the other hand, the input voltage at B is lower than that at E, the "diode" part, E and B, is in a reverse-biased situation corresponding to the second diagram in FIG. 6–7. There is no flow of current between E and C and the switch is off.

A transistor has this switching capability, which a diode does not. In the binary code of all computers—of simple on and off or yes and no—transistors form the basic building blocks of all digital logic units, which then make up the core of all the computers.

Mega-part devices, or integrated circuits Diodes and transistors, as illustrated in FIGS. 6–7 and 6–8, may have created an impression that they are constructed by joining together p- and n-type semiconductors that were fabricated separately. In reality they are made out of a single piece of silicon slab, by blowing in gallium and arsenic vapors onto predesignated areas in many painstaking steps.

The procedures involved in actual manufacturing processes are a lot more complex than explained here, but basically it goes something like this: Take a thin flat piece of a silicon wafer, mask out everywhere except those areas intended for p-type semiconductors, and diffuse in the gallium atoms. Repeat the process for n-types; that is, mask out everywhere, including the p-type areas, except those areas intended for n-types and blow in arsenic vapors. Again, mask out everywhere except where you wish to put in connectors and blow in aluminum vapors on these predesignated paths, having various n- and p-types, diodes, and transistors all connected the way you have designed the circuit.

The number of components that can be crammed into a tiny piece of silicon depends on the ability of the microminiaturization technology used, and this is how a complete electronics circuit containing thousands and millions of transistors can be fabricated within a single small piece of silicon. This is the integrated circuit (IC), more popularly known as the chip.

A crude model of a chip is illustrated in FIG. 6–9, which shows a setup with one transistor, three diodes, three resistors, and several aluminum connectors. It does not correspond to any electronic circuit, but is just something cooked up for the purpose of illustration. You can easily construct, out of a sheet of paper, the three masks needed to spray in, so to speak, gallium, arsenic, and aluminum separately. Compared to today's megachips, it looks like a remnant from the Stone Age, but it contains about the same number of semiconductor components as the world's first integrated circuit, invented back in 1959.

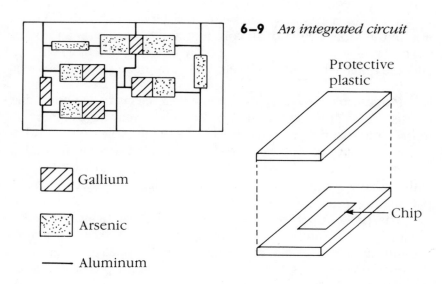

6–9 *An integrated circuit*

Protective plastic

Chip

▨ Gallium

▨ Arsenic

—— Aluminum

We will talk more about chips in the next chapter, but suffice it to say that we have advanced from the days of kilochips to megachips, a memory chip containing as many as 8 million transistors all within the size of about 1 square inch. A rule of thumb says

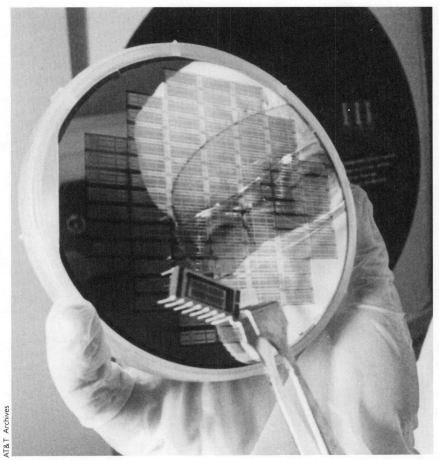

AT&T Archives

6–10 *One of the latest memory chips in a pair of tweezers against the backdrop of a four-inch wafer of doped silicon. A single wafer contains 90 of the megabit memory chips, each chip capable of storing one million bits, or 125,000 bytes or characters, equivalent to about 75 typewritten pages. Some four megabit chips are already available, with 16 megabit chips on the horizon*

that a large-scale integrated circuit (LSI) contains up to 64,000 components, while a very large scale integrated circuit (VLSI) may contain as many as 2 million parts. After a VLSI comes a ultra, ULSI, and even a hyper ultra HULSI.

At this point, you might want to take a brief pause and have a quick look back: integrated circuits out of transistors and diodes; two types of doped semiconductors; artificial mixing of atoms having one more or one less electron than the host atoms; the electron configurations of atoms; the mathematical foundation of quantum physics in the 1920s. From an esoteric theory to the wonder of chips in about 40 years.

7

✷

The core
of a computer

IN THE PREVIOUS CHAPTER we learned about doped semiconductors, how transistors are made out of them, why transistors act as switches, and what an integrated circuit is all about. When a person sits down and faces a computer, does he or she see any one of these items? Not at all. They are all hidden inside a metal box, slightly larger than a typical briefcase, that sits on a tabletop and that we call a personal computer. PCs have become such commonplace appliances—almost as common as televisions or VCRs—that sometimes it is easy for us to forget that they didn't exist prior to 1977.

In and out of the black box

At least three different expressions—*desktop, standalone,* and *personal* describe a more or less identical attribute of a computer. What is meant by these expressions is that a machine is small enough to fit into one corner of a tabletop, is fully self-sufficient, does not need any additional hook-ups, and is directly interactive with an individual user. Typically a PC consists of three separate pieces: a display monitor, a keyboard, and the all-important box. A user hits the keys, looks at whatever is displayed on the monitor, and tries to outwit that darn box—sometimes successfully, but most of the time otherwise. This is a bit of a simplified picture, but it serves to describe the manner in which users interact with a computer.

Most of the time we simply use the computer without particularly caring how it works. A predominant majority, well up to about 90%, of books on personal computers, in fact, deals with instructions for mastering a variety of application programs, including word processing, spreadsheets, databases, graphics, desktop publishing, and other specific professional programs for law, accounting, etc. For a vast majority of the work force, this information is all that is needed. The other computer books provide very useful information on various input and output devices, such as modems, monitors, printers, etc.

This chapter will expand on the knowledge gained from the last chapter and discuss exactly what goes on inside that little black box that sits on a tabletop—the core of a computer. Before we begin to sketch out how the central processing unit of a computer does its thing using its own language of electric pulses, first let us sort things out and separate the black box from the rest of the computer.

All computer hardwares—that is, all electromechanical devices comprising a computer system—can basically be separated into three groups by their functions: input hardware, output hardware, and the black box. Input hardware turns all sorts of input data and graphics into a digital form and feeds them to the black box to be processed. The black box is the "computer" of a computer, inside which all logical decisions and numerical processing are executed at an unbelievably fast speed and following a set of program instructions called an *operating system*. The most important of all so-called software is this set of instructions. The processed digital data are then sent out to a battery of output hardware that converts them into something we can read or recognize.

The task of translating the input and output information to and from digital data is carried out by special electronic circuitries called *boards*. There is a specific board for a specific function. A facsimile board is necessary before you can connect a facsimile machine to your computer; a modem board is needed if you wish to have your computer connected to the rest of the world over the conventional telephone network; and so on. Each input and output device requires a specific board in the computer in order to interact with the computer and do its thing.

A board is so called because the electronic circuits containing scores of chips are mounted together on a single plastic board with preprinted circuits. Sometimes a board is also called a *card*. These boards, or cards, can be inserted into predesigned slots inside the black box; sometimes they are contained inside appropriate input and output hardware. For the purpose of our discussion, let us lump them together with input and output hardware. The division of labor into three groups of a computer system is tabulated in TABLE 7–1 and FIG. 7–1.

TABLE 7–1
Inputs, outputs and blackbox

Input hardwares	Blackbox	Output hardwares
data input	logical decisions, calculations	data output
keyboard mouse scanner fax modem floppy hard disks compact disks	central processing unit	monitor printer plotter fax modem floppy hard disks compact disks

Some devices are strictly input-only or output-only machines. A keyboard and a mouse are input-only hardware, whereas a monitor and a printer are clearly for output purposes only. Most other devices such as data storage devices serve both ends. Magnetic data storage devices such as floppies and hard disks, as well as the more recent vintage of optical storage devices such as video compact disks, serve both functions. The same is true for a modem, which serves to translate digital data into telephone transmission voltages, the term being an acronym of *modulator-dem*odulator!

The global network of telecommunication is gradually being switched over to a totally digital format called the *integrated service digital network* (ISDN), in which voice, data, and graphics can all be sent and received simultaneously in a single digital form. When ISDN becomes the worldwide standard, all modems will have become obsolete, notwithstanding the origin of its name.

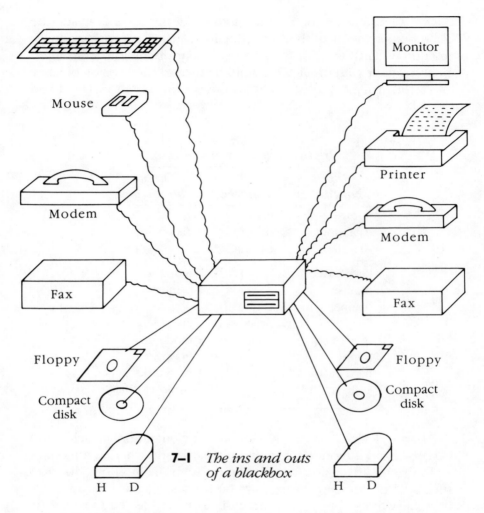

7–1 *The ins and outs of a blackbox*

A facsimile machine is basically a scanner-modem-printer combination. Whatever it scans is turned into a digital signal, transmitted by a modem to another fax, which undigitizes the information and prints up the whole thing.

Up to 70% of the cost of a computer can be taken up by all these input and output devices and accompanying boards. They are, after all, the interfaces that humans use to interact with machines. When you take away all these human-to-machine and machine-to-human interfaces, then and only then you are left with the black box, the "computer" of a computer, which performs nothing short of a miracle using a binary language. It is a two-alphabet language, just a and b—or, equivalently, yes and no.

A two-symbol alphabet

We use a 26-letter alphabet system. A computer does not. We use numerals, punctuation symbols, and signs such as #, ¢, &, and so on. A computer does not. We distinguish between letters and graphics. To a computer, they are merely different designs of graphics. However, a computer makes a big fuss if a colon is replaced by a semicolon.

For a computer, the only language in its universe consists of two symbols, and nothing else. If you reflect on it a little, it is a very simple and primitive system. How would you like to read a book the contents of which are written in their entirety in terms of just two symbols—aababbbababbbabaaaababbbba—! Just imagine the entire volume of literatures in the Library of Congress written in a two-symbol language, not even in terms of a and b, but rather in + and − terms! It is actually not as preposterous as it may first sound. The entire contents of a 24-volume encyclopedia can be stored in this two-symbol language called a *binary code* on both sides (12 volumes each side) of a single optical compact disk only 7 inches in diameter.

A two-symbol language or a binary code can be constructed in as many different forms as there are ways in which to choose a pair of complementary symbols that are easily recognizable and quickly distinguishable from each other. Some of the well-known choices are listed in FIG. 7.2, but you are quite welcome to come up with your own version. How about a horizontal and a vertical line, for example? The long-and-short version shows up more often on mailing envelopes as an automated zip code system. The black-and-white binary code, perhaps the most familiar one, is used as the Universal Product Codes symbol (UPC) for everything from groceries to books and ID cards. About the only places we can still see the dash-dash-dot-dot Morse codes being used are in the scenes of telegraph offices in old western movies!

Then there is the on-and-off version of the binary code, the actual turning on and off of electric currents, the fluctuating pulses inside a computer. This is the language of a machine, an endless sequence of fast pulsing on-and-off of electric signals. To turn a pulse on or off, you need switches—not just a few but a few million. One electric component can act as a fast and reliable switch: a transistor. A black box contains scores of chips, and each of these chips contains as many as a few million transistors. To represent the physical state of the transistors being on or off, we use the 1-and-0 version: 1 for on and 0 for off.

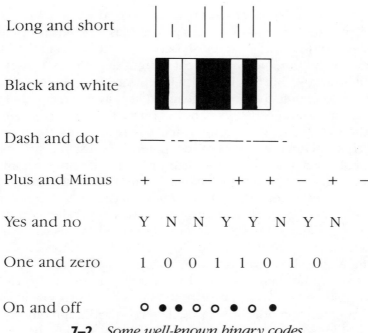

| Long and short | | | | | | | | | |

7–2   *Some well-known binary codes*

In order for us to be able to communicate with the computer, it is necessary to have every input symbol coded into a string of ones and zeroes—everything from alphabets, numerals, mathematical symbols, puncturation symbols, Greek and Korean alphabets, to the black-and-white patterns of your hand-drawn sketch. The only language used by the black box is the actual state of a particular circuit being on or off, depending upon whether a particular transistor passes or blocks an electric current. The two numerals, 0 and 1, are referred to as *bits,* shortened from the term binary digits. In the binary system, these digits take on quite different appearances, as shown in TABLES 7–2 and 7–3.

A zip code, for example, 97146, becomes a little bit longer in binary: 1001 0111 0001 0100 0110. The spacings are inserted just for our convenience, without which it reads 10010111000101000110. We need more than four bits to encode all the symbols mentioned pre-

TABLE 7–2
Numerals in the decimal system

		tens	ones
0	no ones		0
1	one one		1
2	two ones		2
	.		.
	.		.
	.		.
9	nine ones		9
10	one ten and no ones	1	0
12	one ten and two ones	1	2

TABLE 7–3   *Ten digits in the binary code*

		eights	fours	twos	ones	
0	no ones	0	0	0	0	0000
1	one one	0	0	0	1	0001
2	one two and no ones	0	0	1	0	0010
3	one two and one one	0	0	1	1	0011
4	one four, no twos and no ones	0	1	0	0	0100
5	one four, no twos and one one	0	1	0	1	0101
6	one four, one two and no ones	0	1	1	0	0110
7	one four, one two and one one	0	1	1	1	0111
8	one eight, no fours, no twos and no ones	1	0	0	0	1000
9	one eight, no fours, no twos and one one	1	0	0	1	1001

viously. The standard format is to use eight bits, called a *byte,* for every symbol, including some control functions such as carriage return, backspace, delete, and so on.

The standard ASCII system, established by the American Standard Code for Information Interchange, uses 32 bytes for control functions and 96 bytes for all other symbols. In this system, the capital *A* is expressed as 01000001 and the capital *B* as 01000010, whereas it is 01100001 for the lowercase *a* and 01100010 for the lowercase b. The system is quite transparent. The first four bits, 0100, clearly are codes for capital alphabets with A and B being the first, 0001, and the second, 0010, letters. The first four bits for lowercase letters are evidently chosen as 0110.

I am not about to list the full list of bytes for 96 symbols as well as 32 control functions. The important thing is to understand that each single unit of symbols that humans use corresponds to a specific on-and-off pattern of eight switches that a machine recognizes and works with. This is how our language is related to the language of a computer, in the ratio of 1:8. A standard 8½-by-11-inch sheet of paper may contain from 2,000 to 3,000 characters; that is, 2 to 3 kilobytes. One thousand such pages would contain about 2 to 3 megabytes. A 3.5-inch floppy disk comes with 1.2 megabytes of storage capacity, and some hard disks come with capacities of 20, 40, 80, or up to 120 megabytes of space. Now that is a lot of bytes, a lot of typed pages indeed.

The heartbeat of a black box

We now come to the question of just exactly what electronic marvels are contained inside a black box. The answer couldn't be simpler. Inside a black box there are hundreds and even thousands of all sorts of integrated circuits. It is not so much a black box as it is a chip box. The reason for this is obvious. The only real language that a machine speaks is the electronic on and off pulses, and it needs transistors by the millions to perform its function.

These chips can be classified into three principal groups, each performing its own specific task in the division of labor according to specific designs. First, there are the *memory chips,* storing data, information, and instructions either permanently or temporarily. Next is the *central processing unit,* usually called the microprocessor chip. If a black box is the "computer" of a computer system, then the microprocessor is the brain of a "computer" system. Less often mentioned and hence less well known than either the memory or processing chip is the third group of chips, the *clock chips.* This group of chips generates and controls a steady flow of alternating on and off pulses, thereby turning the whole system on and off and keeping every step of calculations synchronized. No part of an operation can get out of step. They maintain a precise heartbeat for the whole computer system. See FIG. 7–3.

Memory chips come in two categories: the ROM bunch and the

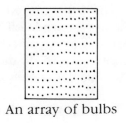

An array of bulbs

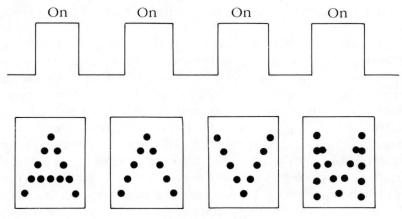

7–3 *A simple program*

RAM clan. The ROM (read-only memory) chips are the permanent depository for two important and crucial sets of instructions. The first is a set of instructions to activate all parts of a computer and get them ready for whatever application they need to perform. When we turn a machine on, all that we are doing is turning on a power source. The instructions stored in ROM do the rest, waking up every part, activating the whole nerve system, and otherwise making sure that everything is in working order. The second set of instructions contained in ROM chips is the all-important software called the *operating system,* which tells the machine how to proceed, one step at a time, with a particular task such as information retrieval, word processing, or any other application program. The instructions stored in ROM chips are permanent. They are always there, ready to spring into action every time we turn a machine on.

The RAM (random-access memory) clan, on the other hand, provides a vast temporary storage space, into and out of which all the data being processed are constantly moved about. A RAM chip goes totally blank when a computer is turned off. It is a vast storage space that you have leased for running a program (FIG. 7-3), and you'd better clear out everything before you turn off the machine because anything you have left in it will be completely erased the moment you turn the machine off.

Data and information are constantly shunted back and forth between RAM chips and the microprocessor, which executes all logical decisions and numerical calculations by controlling the on-and-off gates for the flow of electric pulses through millions of transistors inside it.

In one instant, the black box is turned on and in the next instant the whole thing is turned off, only to repeat the beats all over again. The whole computational unit flickers on and off, so to speak. This may come as a surprise to you, but exactly one-half the time a computer is on, it is off! Each "on" defines one beat or one cycle. Proceeding from one "on" to the next "on" is like going from one frame to the next in a motion picture reel. A moving scene in a motion picture is nothing but a fast display of one frame after another in a film reel. Likewise, each cycle defines a unit of time frame, one frame of a reel, in which the microprocessor proceeds one step in its calculation.

The frequency of pulses, so many on and off in one second, is the basic factor—but not the only factor—that determines the speed of a computer. Every single advertisement of a personal computer touts its speed, but the speed is expressed in terms of the frequency, so many megahertz. This speed refers to the speed of the clock chips that generate the internal pulses. If a particular instruction requires 20 steps, or 20 pulses, for its execution, a machine with a faster speed, a higher frequency rating, will certainly carry out that instruction faster. The unit of one hertz (Hz) is the rate of one cycle per second, so a PC with the speed of 10 megahertz (MHz) operates with an internal pulse that turns on and off 10 million times per second. The time between two successive "on"s is 10^{-7} seconds, or 100 nanoseconds. A motion picture running at this speed would have to show 10 million frames in one second. Even a movie that catches the moment a fired bullet rips into a balloon does not run anywhere near this speed! Some supercomputers operate at the range of a few gigahertz, a few billion cycles per second.

At the risk of an oversimplification, let us compare a black box to an electric display board consisting of rows and columns of light bulbs, such as the simple rectangular one shown in FIG. 7–3—a modest design compared to the gleaming megadisplays in Las Vegas. Suppose that the display board is connected to a pulse clock that turns on and off at the rate of 10 hertz. One of the simple programs can be something like a four-cycle display that changes a letter *A* to *M*. At the first "on", the first cycle, the board displays the pattern of an *A*. Momentarily, the display goes off. At the second "on", an instruction turns on those bulbs for an *A*, except the horizontal bar. The third instruction is a little bit fancy. It turns the pattern upside down, by lighting up correct bulbs in each column. Again, momentarily, the

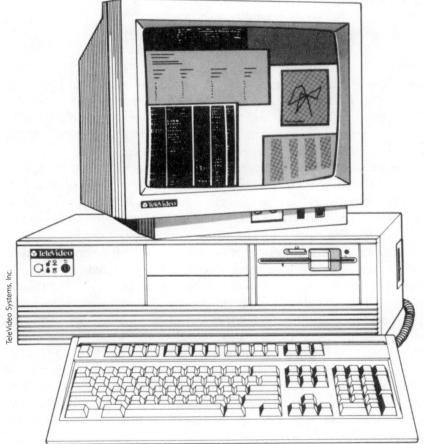

TeleVideo Systems, Inc.

7–4 *A standard assembly of a personal computer system. The keyboard (input), the display (output) and the blackbox*

whole thing is turned off, and at the fourth cycle, you can guess what the instruction should be. It proceeds this way, cycle by cycle, frame by frame.

Crude as it may be, this example serves to illustrate, rather accurately, what goes on inside the black box among the clock, memory, and microprocessor chips. Data processing proceeds just like it, one cycle to the next, one frame at a time, byte by byte, and through a complex network of millions of transistors. An operation of a computer, at the very heart of it, can be summarized in one sentence as follows. A large number of tiny semiconductor devices, transistors, work remarkably fast, repeating a very simple act of opening and closing switches an incredible number of times! That is all that happens inside the black box of a computer.

Microprocessors

Not a single advertisement of personal computers fails to boast the speed of its microprocessor. "At 10 megahertz, it runs your programs 30% faster than . . . . . .", ". . . up to four times faster, based on the 80286 microprocessor," or "the incredible power of an 80386 microprocessor operating at a blinding speed of 25 megahertz . . ."

How is the speed of a computer or, more specifically, the speed of a microprocessor determined? Is a computer with the speed of 16 MHz twice as fast as the one chugging along at 8 Mhz? Not necessarily. Usually it is less than twice as fast. The advertised speeds refer to the raw speed of the clock chips we discussed in the last section.

The brain of a computer, where all nemerical calculations and logical decisions are executed, is the CPU. For a mainframe or a supercomputer, a CPU is quite sizable in its physical dimensions (FIG. 7-5). The CPU alone for a relatively low-end supercomputer costs close to a few million dollars, for example. For smaller desktop computers, personal computers, and workstations, however, the entire circuitry of a CPU can be crammed into a single chip. This CPU-on-a-chip is what is called a *microprocessor,* and a computer whose CPU is a microprocessor is called microcomputers. Because most of us deal only with desktop computers and rarely with mainframes, the two descriptions, CPU and microprocessor, are quite often used interchangeably.

The computational speed of a computer—the performance that really counts—is determined by three principal factors, only one of which is the raw speed of the clock chips.

At the risk of an oversimplification, let us compare a microprocessor to an automobile engine. The raw clock speed can be equated with the speed of rotation of the crankshaft. The number of revolutions per minute of the shaft is measured by a device called a *tachometer.* At 1200 rpm, the engine is running at the rate of 20 revolutions per second, each cycle corresponding to a time interval of one-twentieth second. In a computer, as mentioned before, a clock

Photo: Paul Shambroom, Cray Research, Inc.

7–5 *A four-processor Cray supercomputer system. In order to achieve much faster speed, several processors divide up a complicated program into many parts and compute simultaneously. This technique, known as a parallel processing, is what separates a supercomputer from all other computers including mainframe computers*

rate of 20 MHz defines one cycle to be a slice of time corresponding to 40 nanoseconds.

The power of an engine also depends on the number of cylinders—4 cylinders, V-6, V-8, and so on. In a similar manner, the power of a microprocessor depends on what is called the *bit size,* which tells us how many bits, and hence how many bytes, the microprocessor can access instantaneously. Since each bit entertains only two possibilities, either one or zero, a microprocessor with a 4-, 8-, 16- or 32-bit size has instant access to the number of bits equal to 2 multiplied 4, 8, 16 or 32 times, which correspond to 16, 256, 65,536 and 4,294,967,296 bits respectively. Almost all microprocessors today operate with a 32-bit size; that is, the computer has instant access to 4 billion or so bits to retrieve, calculate, or otherwise process (see FIG. 7-6.).

In addition to the bit size and the clock speed—that is, the number of cylinders and the speed of revolution—there is a third factor that determines the speed of a microprocessor. This factor has to do with the question of how many cycles, on the average, it takes for one instruction of a program to be completed. This aspect of the speed depends more on the internal design of a black box than just the raw speed. Just as the instruction of changing *A* to *M* took four cycles in the previous example, one instruction in a program, says, "if such is such and that, then go to here," requires quite a few cycles to complete all the steps involved.

This number of cycles to an instruction can be compared to a transmission, which converts so many rotations of a crankshaft to a single rotation of a drive shaft, which then turns the wheel. Let us take one of the simplest "computers," a digital watch. Typically, a tiny quartz crystal vibrates at the rate of 32,768 times per second, generating pulses at 32.768 kilohertz. The coded instruction to advance one second on the display—say from 11:15:06 to 11:15:07—is executed only after 32,768 cycles!

The first mass-produced commodity microprocessor was a 4-bit chip named 4004, produced by Intel Inc. in 1971. In 1974, Intel manufactured the first 8-bit microprocessor, 8080, which was capable of adding two numbers in three-millionths of-a second.

Two companies, Intel and Motorola, are the Coca Cola and Pepsi Cola of the microprocessor business. They dominate the microprocessor market, even though others are catching up rapidly. Almost all personal computers, with the notable exception of Apple products, predominantly use Intel chips, while the situation is exactly

AT & T Archives

7–6   *One of the first 32-bit microprocessors developed by Bell Labo-*
ratories. The dime size chip contains about 150,000 transistors,
so many different patterns of doping.

reversed for the engineering-intensive workstations, which rely heavily on the Motorola chips. The four most widely used microprocessors are listed in TABLE 7–4.

The performance speed of a computer is often expressed in terms of the number of instructions it can execute per second. This speed is an average only, since different instructions require a different number of cycles. One MIPS stands for 1 million instructions per second, a speed of a few MIPS being the standard for a state-of-the-art 80386-based machine. Another expression often used is what is called a *flop,* a floating point operation per second. Some of the fastest machines at present boast the range of a few gigaflops—1 billion floating point operations per second. Insofar. as one floating point operation is just one particular, albeit more mathematical, instruction, one megaflop and one MIPS refer to about the same speed.

TABLE 7–4 *Some recent microprocessors*

Year introduced	Microprocessor	Speed in megahertz	Bit size	Average number of cycles per instruction
1983	Intel 80286	8	16	6
1984	Motorola 68020	12	32	7
1986	Intel 80386	18	32	4
1987	Motorola 68030	18	32	5

8

✳

Trading energies between light and atoms

JUST AS THE KNOWLEDGE of atomic structures enabled us to develop the technologies of doped semiconductors and integrated circuits, an understanding of the intricate mechanisms by which atoms and light trade energies between them was to spawn another marvel of today's high-tech, a precisely aligned beam of radiation generally called a *laser light*.

Absorption and emission of energies

Before you can understand a laser light and how it is produced, it is necessary to come to know the mechanisms by which an atom absorbs and emits energies in the form of radiation, a subject usually known in the trade by a rather imposing title: the interaction of the electromagnetic radiation with matter.

An atom can absorb one color and turn out a different color, literally. It is nature's own frequency converter, and we draw benefits from it every time we turn on a computer display, a color television or a fluorescent bulb, all of which are not even lasers. A laser is one of those words that can be understood best by spelling it backward one letter at a time:

r, radiation

er, emission of radiation

ser, stimulated emission of radiation

aser,   amplification by stimulated emission of radiation

laser,   light amplification by stimulated emission of radiation.

We have already discussed radiation in chapter 3. The emission and absorption of radiation by atoms, as well as a special class of emission called the *stimulated emission,* are subjects that are as old as quantum physics itself. In fact, it was through the processes of absorption and emission that we came to learn of the energy levels of atoms, and the theory of quantum physics was invented to explain them in the early part of this century. It wasn't until 1960, however, when a trick was discovered to help sustain stimulated emission and a laser was invented.

We will defer the discussion of a laser light until the next chapter. In this chapter, let us deal with the basic aspects of how an atom gulps up and coughs out electromagnetic energies.

In their natural state, atoms always stay at their lowest possible energy form, the stable electron configuration called the *ground state.* All the configurations given in TABLE 4–3 in chapter 4 correspond to the ground states of each atom. Now, when and if a correct amount of energy is supplied to an electron by an electromagnetic radiation such as a light, the electron can and indeed does absorb that energy, and promotes itself to a state of higher energy—higher by the exact amount that it has absorbed. Three examples of such transitions from a lower to a higher energy level are given in FIG. 8–1. For brevity, only shells are indicated, suppressing subshells.

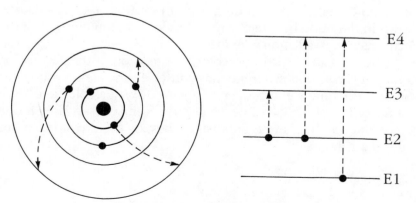

8–1 *Examples of absorption*

In the process of absorbing energy, or the transition to a higher state, an electron is extremely fussy. The energy supplied must match exactly the difference between the two levels involved—neither a bit less nor a bit more. If the energy supplied is either a little more or a little less, it will be totally rejected by the involved electron, for a very simple reason. The electron has no place, a level, to go to. It cannot exist in any other energy state than those levels allowed by quantum theory. In the same way, we cannot hold onto a missing rung in a ladder.

Once up, an electron must take advantage of the height and look around very quickly because it can stay there for a fraction of a fraction of a second—about 10^{-8} second on the average. It usually has to come right back down, back to the good old stable, safe, and comfortable ground state. On its way down, an electron gives back the energy it acquired during the absorption. Now it emits radiation. It must shed itself of exactly the same amount of energy that was put in.

In the emission process, however, it has more than one option. An example of the options an electron has in coming down to the ground state from a higher state is shown in FIG. 8–2. A single amount of energy corresponding to the difference, $E_4 - E_1$, was needed in going from level one to level four in one jump. Of course, an electron can go up in steps also, but it usually makes a transition down to a lower level before it can be stepped up to the next higher level.

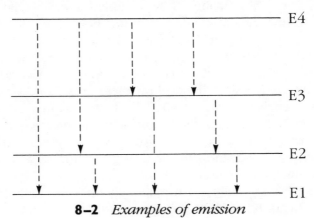

8–2 *Examples of emission*

Not that we need another weird example, but suppose we have this strange thing called a "quantum" dough that is being baked in a microwave oven. At some hot temperature, it suddenly doubles in size, and at a still higher temperature it abruptly triples in size. As the bread cools on a table, it suddenly shrinks back to its original size and emits microwave radiation right on your face!

Photons

As atoms absorb and emit radiation energy, they do it in strict accordance with the level structures of their electrons. Only an exactly matching amount of energy can be absorbed or emitted when an electron makes a transition between levels.

Up to this point, however, we have not really addressed the question of just how the energy of a radiation can be related to its other attributes, such as wavelength or frequency. We know that various portions of the spectrum have different energies, an x-ray clearly having more energy than a visible light, and "a sunshine on our shoulder makes us happy" much more so than any infrared beams. Evidently the energy of a radiation cannot depend on its speed, because all radiations have one and the same speed.

The properties of the energies of a radiation at the level of atomic physics–that is, how they come in an indivisible quantized unit to be emitted or absorbed,–form a topic that is not exactly an easy subject to come to terms with. It is a somewhat subjective exercise to say which topic is more difficult than others, but back in chapter 4 I proclaimed that chapter to be perhaps the most difficult part in this book. The subject matter of this chapter can easily qualify as perhaps the second most difficult part of this book–the difficulty being more conceptual this time.

Referring to the electromagnetic spectrum, as shown in FIG. 3–5, the portion of radiation we are concerned with in the world of atoms includes infrared, visible light, ultraviolet, x-rays, and gamma rays–that is, frequencies higher than 10^{13} hertz. In this range the electromagnetic radiation begins to exhibit quantum nature, which is not really all that surprising because it gets absorbed or emitted by atoms whose energies are quantized. The quantum nature of a radiation has the following two basic aspects.

First, for every different frequency within the range just mentioned, a radiation has its own characteristic energy unit, an indivisible basic lump of energy that is to the energy of a radiation what pennies are to a currency system. The basic energy unit for a red light is different in magnitude from the basic energy unit for a blue light, which is different from that of a soft x-ray, and so on. Pennies define the basic unit of money, but U.S. pennies are different from Canadian or German pennies in value. A beam of a green light may contain billions of billions of the green energy units, effectively washing out the discreteness, but in the final analysis, at the atomic scale, its energy is a countable sum of its own green energy units.

This can be likened to a vast expanse of a beautiful sand beach. The beach can appear as a continuous creamy spread of sand from a distance, but upon a closer examination it is made up of trillions and trillions of individual and countable sand granules. Each sand granule is a quantum of a sand beach. Sand granules from a South Carolina beach may be finer, smaller in size, than those from the rocky shore of Maine.

Such quanta of radiation are called *photons*—a green photon for a green light, a violet photon for a violet light, a hard x-ray photon for a hard x-ray, and so on.

The second aspect is a quantitative one. The energy of a photon of a radiation is directly proportional to its frequency, and this explains the earlier remark that an atom is nature's own frequency converter. As an atom absorbs one amount of energy and then emits two smaller amounts of energy (FIG. 8–2), it is absorbing a photon of a higher frequency and emitting two photons of different and lower frequencies. From FIG. 3–5 let us pick three frequencies—for example, a hard x-ray with the frequency of 10^{20} hertz, a soft x-ray with 10^{17} hertz, and a red light with 10^{14} hertz—their frequencies being in the ratio of 1 million to 1,000 to 1. The energy of a single hard x-ray photon is 1,000 times that of a single soft x-ray photon, which in turn is 1,000 times that of a single red photon.

Compared to human-sized dimensions, these photon energies are insignificantly minuscule, but the disparity and spread in their magnitudes are quite large. With respect to the photon of a radiation with the frequency of 10^{12} hertz, for example, a single photon of a gamma ray of 10^{22} hertz has an energy that is 10 billion times that of the former.

In FIG. 8–3 we have enlarged the portion of the electromagnetic spectrum that contains visible light. Along the horizontal line are marked the frequencies between 10^{14} and 10^{15} hertz, and photon energies are plotted vertically in a totally arbitrary scale, just to indicate the relative magnitudes only. A single photon of a radiation is either absorbed or emitted by a single atom in one transition. Billions of billions of atoms emitting billions of billions of photons, up and down a range of frequencies, are the sources of the light that we see on televisions, fluorescent bulbs, lasers, and the like. As shown in FIG.

8–3, a photon of an ultraviolet ray of ten units of frequency has an energy equal to ten times that of a photon for an infrared having one unit of frequency. The spread in frequencies in the visible spectrum covers a narrow range from about 4.5 to 7.5 units in 10^{14} hertz.

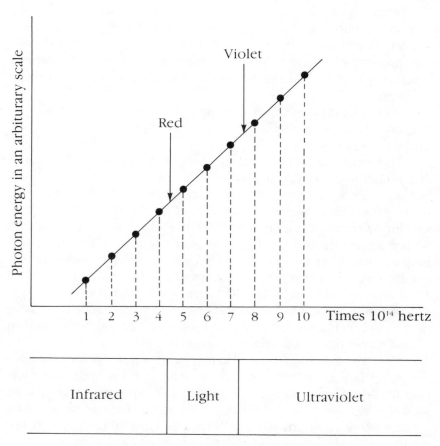

8–3 *Photon energies as a function of frequencies*

Fluorescence and phosphorescence

Let us now consider a particular combination of absorption and emission in which a single transition is made to a higher level, followed by a two-step transition back to the original level, all within the frequency range shown in FIG. 8–4. The absorption enables an electron to go from level E_1 to, say, level E_{10}. In order for this to happen, an

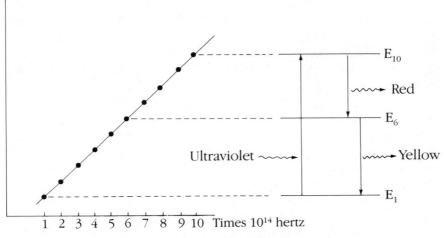

8-4 *Fluorescence*

electron will have to absorb a photon having exactly nine units of energy, an ultraviolet ray of frequency 9 times 10^{14} hertz. No other radiation, no other frequency, will do. Suppose the electron drops back down in a two-step emission, first to E_6 level and then to E_1. The atom is emitting two photons: one with frequency 4 times 10^{14} hertz, a red photon, and the other with frequency 5 times 10^{14} hertz, a yellow photon. This is what happens with just one atom.

The situation just described corresponds to the case in which an ultraviolet (UV) ray produces a combination of red and yellow lights, a process called *fluorescence*. Inside a fluorescent lightbulb, electrons accelerating from one end to the other end of a tube, when it is turned on, collide with the atoms of gas filler producing ultraviolet rays. These UV rays are absorbed by atoms of materials coated on the interior wall of a bulb and we get the illumination we need as the result of a simple fluorescence: a one-step absorption followed by a two-step emission.

Phosphorescence involves a mechanism that is a little different from, though essentially similar to, fluorescence—a mechanism that gets us one step closer to a laser beam of light. This mechanism involves the question of how long a time, or how short a time, can an electron stay around in a higher energy level. Except for the stable ground state, every other energy level of an atom has a characteristic lifetime of its own, some allowing an electron only a very short visit; others, a relatively long time.

Not only are they different from each other, but these lifetimes are not all that precise. A typical range of lifetime before emission is between 10^{-7} to abut 10^{-9} second—between 100 nanoseconds to 1 nanosecond—some much shorter and others much longer. Within this range of lifetime, an electron drops to a lower state anytime it cares to.

There are some energy levels whose lifetimes are much longer than normal, allowing electrons to park and stay up to 10^{-3} second— one-thousandth of a second—and some even up to several seconds and longer! This is an unbelievably long time by atomic standards. Once elevated to such a *metastable state,* as these levels are called, electrons will take their sweet time before emitting photons. Such a delayed emission is called *phosphorescence.* We are all too familiar with it. We observe it every time we look at that green glow in the dark telling us that it is time to get up and go to work. All luminous dials are made with such phosphorus stuff.

A typical two-step process for a phosphorescence is shown in FIG. 8–5, in which a visible light is absorbed to go up to a higher level. The electron immediately drops to a lower level emitting an invisible infrared and after a while drops back down to the level where it started, with an accompanying delayed emission.

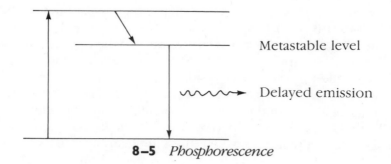

8–5 *Phosphorescence*

Stimulated emission

In a general emission process, an electron drops to a lower level when it is good and ready (within the range of the allowed lifetime of each level) at different times, some sooner and others much later, but at the time of its own choosing nevertheless. Now there exists an entirely different mechanism for an emission, called the *stimulated emission,* in which an electron elevated to a higher level is prodded, induced, or stimulated by a passer-by photon to drop to a lower level before its time,—a passer-by photon inducing an early labor, so to speak.

Isolated incidents of stimulated emission always accompany a general emission process, and a mathematical analysis of a stimulated emission was put forward as far back as 1916 by none other than Albert Einstein. Under normal circumstances, what little stimulated emissions that occur die out very quickly in a medium and cannot be sustained continuously to be of any particular use. It was to take about 40 years since the initial theory was put forth until a technique was invented to sustain a chain reaction of stimulated emissions, to wit, what we now call a *laser light.* Let's see now how this process comes about. It is just another way in which a photon and an electron play together.

Take two identical atoms moments after each has absorbed a photon from a radiation, as shown in FIG. 8–6. Atom A is ready for emission, while atom B would like to hang around in an excited higher state for a moment or two longer. The electron of atom A says, "Well, I don't know about you, B, but I have seen enough and I am about ready to go back down," and jumps down emitting a photon. This photon, if it is traveling toward B, can hit the electron of atom B, while still up there at E_2 level, and gently nudge and talk it into dropping down before its time. The result is a pair of identical photons coming off: the one from atom A who is doing the convincing and the second photon, identical to the first, emitted by atom B. They move in the same direction and both have the same energy, and since the stimulation by the first photon and the emission of the second photon are simultaneous, they both correspond to an identical wave pattern—two crests match and likewise two troughs match. If the two photons run into other atoms in their ground states, they will just be gobbled up and that is the end of one particular event of a stimulated emission. Since, under normal circumstances, atoms in their ground

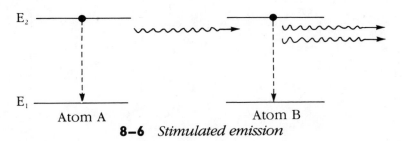

8–6 *Stimulated emission*

states far outnumber any in a higher state, stimulated emissions cannot normally be sustained.

Now this might be a little bit corny, but let us compare this situation with ripe apples falling off a branch of an apple tree. In one case, a ripe apple falls to the ground when it is good and ready. You might say that this corresponds to an ordinary emission. In the second case (FIG. 8–7), an apple is thrown up from the ground to gently touch and disturb an apple still on a tree. The one that is being thrown is the inducer, and it must be neither underthrown nor overthrown. You've got to do this with the right touch so that the inducer reaches its top height at the location of the apple on a tree. Then both apples will come falling down. They travel in the same direction, have the same energy, and are in perfect unison in their motion.

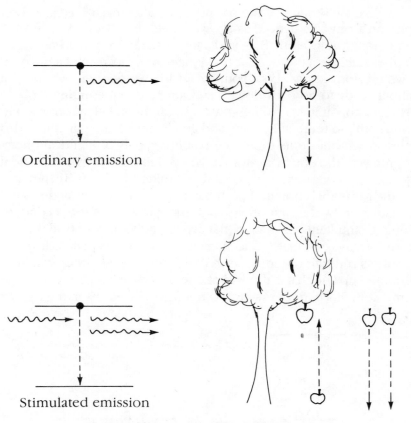

Ordinary emission

Stimulated emission

8–7 *Falling apples, the eating kind*

A stimulated emission of a photon is exactly like that. In principle, you can now throw up two apples and four of them come down in unison, four into eight, and so on. You would have a shower of apples coming down in unison. That is the very idea of a laser light: a shower of photons in unison.

9

＊

Lasers

APPLICATION OF LASER TECHNOLOGY covers a wide range of power, as well as sophistication, from a check-out counter of a modern supermarket to an antimissile defense system popularly know as the Star Wars program. Flickering patterns of zapping beams of laser lights also became an indispensable part of a rock concert. It adds dynamics! Most of these applications of laser technology are relatively recent—1970s and 1980s—even though the principle of stimulated emission had been known since 1915. It wasn't until 1960 when the first laser was invented.

A chain-reaction of photons

As discussed in the last chapter, a stimulated emission doubles the number of photons per participating atom—a sort of a cloning factory of photons having the same energy and moving in the same direction perfectly in step. If we can sustain a continuous chain reaction of this process, we would end up with a very sharp and powerful beam of light. Such a chain reaction of stimulated emissions is what is referred to as the *amplification by stimulated emission of radiation,* the ASER for short. An ASER done with light is simply a laser.

After having understood how a stimulated emission works and how phosphorescence takes advantage of electrons staying longer in a metastable state, we can see very easily how to get an ASER mechanism going. The recipe is simple. Take a sample of atoms (a gas or a

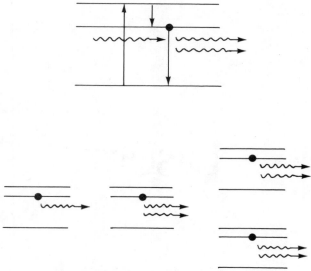

9–1   *Amplification by stimulated emission*

piece of a solid), shine radiation onto it (a light, an ultraviolet, or even an x-ray), and by a process similar to phosphorescence get as many atoms as possible, certainly a majority of them if not all, elevated to suitable metastable states. That is absolutely all that we have to do. We can just relax, sit back, and let nature do its thing.

The top part of FIG. 9–1 illustrates a single atom. As in the case of phosphorescence, an initial absorption of an externally supplied radiation energy—called an *optical pumping* in the trade, for an obvious reason—raises the energy of an electron from a ground state to a third level, from where the electron almost immediately drops to the second level, which we choose to be a metastable state. The lifetime of an electron in the metastable state need not be as long as in the case of phosphorescence, but just long enough for it to hang around until a passer-by photon comes knocking. If we have a sufficient number of atoms momentarily in their metastable states, we are all set for the lasing action. Sooner or later, one of the electrons in a metastable state will look round and say, "OK, you guys, I am going down." We don't have to do a thing, since there will always be the first electron.

As shown in the bottom part of FIG. 9–1, the first photon triggers a stimulated emission in the neighboring atom, producing two photons: itself and its clone. Since there are plenty of atoms still in their metastable states, the two photons trigger two more stimulated emissions, producing four photons, including two more clones. Now you don't know and care which is the original and which are the clones! Off they go, four into eight, eight into sixteen, and so on until you have a cascade of photons, all having the same energy and hence the same frequency—pure red, pure blue, or pure green—and all aimed in the same direction. The result is a narrow, very sharp, pinpoint accurate, and very powerful beam of light, which we have come to call a *laser light*. A production of a laser light is really a simple matter.

A stream of photons—a beam of light—goes through repeated reflections at both ends of a material (FIG. 9–2), increasing the number of photons attaining a higher intensity, and after having gained a sufficient intensity to break through, emerging through the partially transparent small opening on one side. Because the beam is of only one frequency, in a visible range, it corresponds to a single color: a red, a cold blue, or a dark violet laser light, flickering on and off, forming geometrical patterns, in tune with the ear-shattering and bone-vibrating beats of a rock band. Stimulated emissions can, at different times, stimulate more than just a lot of photons!

A familiar red laser beam is produced by a ruby-rod laser, as sketched in FIG. 9–2. A small fraction of aluminum atoms in this

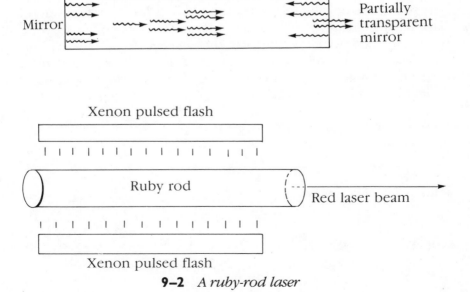

9–2 *A ruby-rod laser*

aluminum oxide compound is replaced by chromium impurities, and it is these chromium impurities that are doing the "lasing". A xenon pulsed-flash system is the energy pumper, raising electrons to their metastable levels. Each time the flash pulses, we get a red laser beam. A ruby laser produces a red laser beam of the wavelength in the range of 0.7 microns—seven-tenths of a millionth of a meter. It has about the longest wavelength, and hence the lowest frequency and energy, in the visible light spectrum. It is most commonly used in a simple scanning purpose such as a reading of bar codes, the Universal Product Codes on grocery items.

Masers, lasers, and "xasers"

A stimulated emission clearly requires a physical system that is made up of certain constituents and whose energy levels are discrete, so as to allow absorption and emission mechanisms to operate. We have based our discussion, up to this point, on the atomic structures, but there are other physical systems that can sustain a similar chain reaction of photons, molecules and atomic nuclei in particular. In the case of molecules, the energies involved are much lower than those of optical spectra, whereas the energies absorbed or emitted by the discrete energy levels of protons and neutrons inside a nucleus are several orders of magnitude larger than what come out of atomic emissions. Historically, the first accomplished "lasing" action was, in fact, obtained using molecules, in 1954, and it took another six years before the first atomic "lasing" action was achieved.

The energies released in stimulated emission processes from these three systems cover a wide range of the electromagnetic spectrum from a microwave to a hard x-ray, as summarized in FIG. 9–3.

Sources of emission	Types of radiation	Typical frequency range in hertz	
Molecules	Microwave	10^{10}	
Atoms	Light	$10^{14} - 10^{15}$	
Nuclei	X-rays	10^{20}	

9–3 *Lasers from different sources*

This is perhaps a good place to make another comment on the art of terminologies in physics. The ASER process obtained from molecules produces a sharp microwave beam, and the beam is appropriately called a *maser*. The light ASER process produces a sharp optical beam, and it is called a laser. But when a sharp x-ray beam is produced by an ASER process off an atomic nucleus, it is called an x-ray laser. Strictly speaking, it should have been called a *xaser*! You see, an x-ray is not even a light. Of course, it might have been somewhat interesting if an engineer from Xerox Corporation had invented a totally new x-ray laser printer. A xaser printer from Xerox?

Anyway, the name *laser* has outgrown its shell and now it stands for any beam of radiation that has a single frequency and moves in the same direction in perfect unison. Principal differences between an ordinary light and a laser light are summarized in TABLE 9–1.

TABLE 9–1 *An ordinary and laser light*

	Ordinary light	*Laser light*
Frequency	All frequencies in the visible spectrum	One frequency, one color
Coherence	Rays out of steps with each other, random steps	All waves in steps, crest to crest, trough to trough
Resolution	Wide beam spread, diffuses wide	Very little beam spread, pinpoint and sharp
Power	Up to a few kilowatt	Up to a gigawatt, even a terawatt
Uses	Illumination	Precision and /or power tool

The first entry in the table concerns the frequency of the two kinds of light. An ordinary light, the white light, contains the full range of frequencies, from the reddest red to the darkest violet, all jumbled up. Even in the case of fluorescent bulbs made up of only two or sometimes four or more frequencies—depending on the nature of gas mixtures used inside a tube as well as the interior coating materials—we get slightly different tints of a white light, some more bluish and others more pinkish, since the component waves are all randomly out of step. A laser light is of one frequency only. It is strictly *monochromatic,* of one color.

The second entry in TABLE 9–1 is what is called a *coherence.* This term has nothing to do with whether or not a beam of light makes sense or is intelligible to us. When two or more waves are in perfect alignment with each other, fluctuating up and down in steps, they are said to be *in phase.* A group of waves having all the same frequency and in phase is called a beam of *coherent light.*

Coherence is one of the key properties of a laser light. In an ordinary emission, electrons elect to drop down to a lower level, emitting photons completely randomly within an allowed range, and there is no way of attaining any degree of coherence under these circumstances. Furthermore, in an ordinary emission, photons are emitted every which way, whereas in a stimulated emission, the two photons—the stimulator and its clone—come out moving in the same direction.

Thus, an ordinary light is a totally incoherent collection of many frequencies spreading out in all directions, whereas a laser light is a monochromatic and coherent beam of light traveling in one direction with almost negligible spread. A beam of laser light was bounced off a mirror left on the moon by astronauts just for this purpose. These properties are shown in FIG. 9–4.

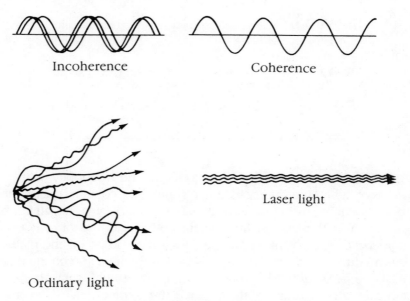

Incoherence   Coherence

Laser light

Ordinary light

9–4 *Ordinary and laser light*

The rate of power consumption for household lightbulbs is usually between 60 watts and 200 watts—300 watts at most. Stadium night lights have up to a few kilowatts per lamp. You can size up the power of some of the laser lights by looking at their frequencies. A single x-ray photon for an x-ray of frequency 10^{20} hertz packs an energy that is 1 million times the energy of a photon for a light of frequency 10^{14} hertz. At a relatively lower energy, a laser light can be used for precision surgery, and at the higher end—megawatts, gigawatts, and even going upscale to a terawatt range—it can be employed as a metal-piercing weapon. One amazing aspect of laser technology is that such a powerful beam of laser light can still be reflected at all by some special optical devices!

Applications of laser technology are developed along two main characteristics: its precision, and/or its power. Applications that depend on a laser's precision and not so much on its power include: pinpoint accuracy in measuring distances and angles, alignment over a large distance, medical surgeries, rifle gunsights, laser guidance for bombs and missiles, as well as more familiar instruments such as laser scanners, laser printers, laser facsimile machines, laser compact disks, and laser video disks. Those applications that utilize both precision and power include: industrial cutting and welding tools, the antiballistic missile defense system, as well as laser-induced superhot heat for a possible nuclear fusion research. (Now every laser requires an initial energy input, the absorption to a higher state. In the case of an x-ray laser, this initial input requires a small nuclear explosion.)

Optical fibers and lasers

The subject of optical fibers is not by itself related to any mechanism generating a laser light, but laser pulses are transmitted through these fibers, opening up an entirely new field of technology known as *optoelectronics*. What is surprising about the relatively recent entry of optical fibers into the family of today's high-tech is that it took as long as it did, especially since it depends on a relatively simple application of a physical phenomenon as old as optics itself.

The phenomena of reflection and refraction are so basic they practically serve to define what a light is. The speed of light is usually quoted in terms of its value in a vacuum, but in traversing a material medium its speed does vary a little because of the electric and magnetic properties of the medium. It slows a bit going through a dense medium, and this is what causes the phenomenon called *refraction,* or the bending of light, at an interface between two different media (FIG. 9-5). This, of course, is what makes a straw look bent in a glass filled with water and adds a little excitement in fishing, a fish being actually nearer than it appears.

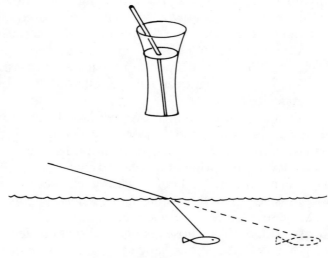

9–5 *Refraction of light*

In our daily experiences, we almost always look from air into denser materials such as water or quartz, but rarely the other way around, and we do not run into many opportunities to experience at firsthand a special case of refraction called *total internal reflection*. Just as a ray of light bends "in" when going from air into water, it will bend "out" when coming into air from water. As you look up from the bottom of a pond, objects appear to be nearer than they are, as would be the case for rays A and B shown in FIG. 9–6.

Ray C in FIG. 9–6 is so refracted that it does not really leave water, bent to skim the surface. The angle ray C makes with respect to a vertical is called the *critical angle*. An object seen through ray C would appear rather near while in actuality it might be an infinite distance away.

A ray of light hitting the interface with a slant greater than the critical angle, such as ray D in the figure, is so bent at the interface that it cannot leave the water at all. It is all reflected back into water. This phenomenon of refraction beyond the critical angle, reflecting all rays back into its original medium, is what is called *total internal*

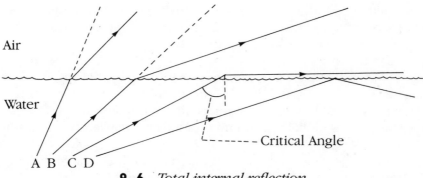

9–6 *Total internal reflection*

reflection. Nothing escapes; all is kept within. Even though it sounds as if it is implying something closely related, a total internal reflection is not in any way related to the practice of Zen meditation!

This old phenomenon of total internal reflection lends itself to a convenient technology of containing and transmitting light signals, including coded pulses of laser lights, through a tube or a plastic wire, which is not altogether all that difficult to manufacture. Such light-containing and signal-transmitting tubes are aptly called *optical fibers* (FIG. 9-7). You have seen them in a lamp store or in a hospital. Crudely speaking, an optical fiber is a solid tube of light-transmitting material, wrapped by a concentric hose made from a material that is less dense than the tube. In practice, a lot more layers are involved for the purposes of protection, and some amplification along the line.

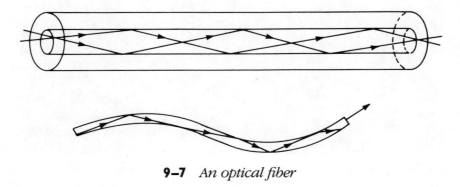

9–7 *An optical fiber*

One of the most familiar and perhaps the least sophisticated use of optical fibers occurs when a bundle of plastic light pipes is formed into a shape of a willow tree, making an interesting conversation piece sitting on an end table in your living room. In medicine, an endoscope made similarly with a bundle of light pipes enables a physician to illuminate and see your innards.

The advantages of optical fibers are manifold. They are free of electric and magnetic interferences since they are not metals; they can last a lot longer; and, more importantly, they can carry substantially more digital information than conventional metal wires. Some local as well as global telecommunications are carried by optical fibers.

Now a laser light can be transmitted through optical fibers also. A string of binary on-and off electrical signals can be converted to a string of on-and-off laser pulses. These signals are then transmitted over a network of optical fibers, and at the other end the laser signals are converted back into electrical signals. What we have here is a fastly developing technology that brings together our state-of-the-art knowledges and technologies of semiconductors, lasers, and optical fibers. It is the technology of *micro-optoelectronics.*

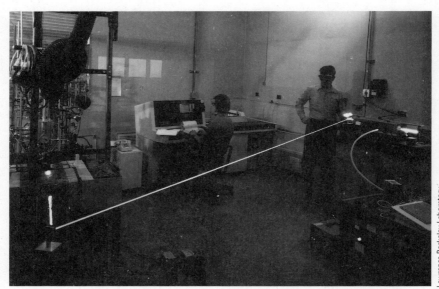

9–8 *A tunable laser, used for chemical research, in operation*

AT&T Archives

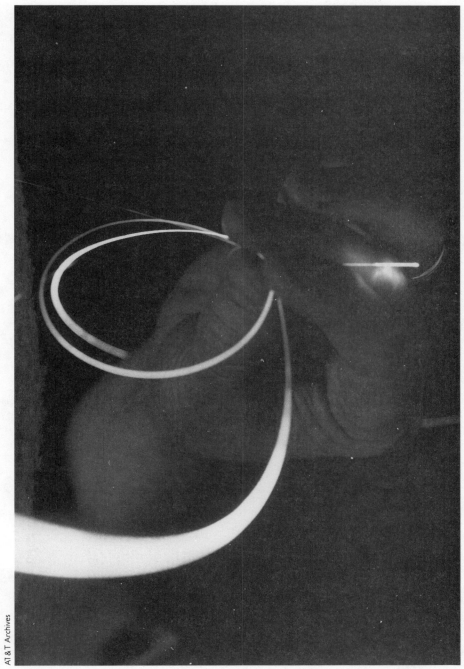

9–9 *Loops of a hair-thin optical fiber made of glass, illuminated by a laser light*

10

✳

Superconductors

IN THIS CHAPTER, we will deal with a subject matter that is as exotic as it is cold. The physical phenomenon called superconductivity is something we can only describe, but cannot yet explain, because as of today it is not yet understood.

The old and the new

The history of superconductivity, some 80 years of it, is remarkable for its wide time gaps between significant discoveries, of which there have been four milestones. The first was the original discovery itself in 1911. Then came the discovery in 1933, a full 22 years later, of its peculiar magnetic property, in which it expels all traces of magnetism from its interior, called the *Meissner effect*. The first, and so far the only, theoretical understanding came in 1957 in the form of the so-called BCS theory, named for John Bardeen, Leon Cooper, and Robert Schrieffer. It is a good theory, but it cannot explain the new kind of recently discovered superconductor. Another 29 years were to pass, a full 75 years after the initial discovery, before a sudden and dramatic announcement in 1986 of the discovery of a ceramic superconductor, of all things. How long will it be before we can come to a comprehensive understanding of all this? Another 46 years?

It is not that the subject is singularly difficult, as these time gaps would at first seem to suggest, but rather that the whole field has been suffering from chronic inattention. The year 1911 was right smack in the middle of the miracle period, in which the foundations for the revolutionary modern physics were being laid, and most people were just too occupied with other things—the relativity of space and time, quantum theory, atomic and molecular physics, nuclear structure, solid-state and elementary particle physics, cosmology, and so on to give superconductivity much thought. This situation has all changed since 1986. Now it is one of the hottest areas of research and development among the industrialized nations.

In talking about the supercold world of superconductors, we often hear about the absolute zero temperature; that is, zero on an absolute temperature scale called *Kelvin*. Of the three standard temperature scales, two are a matter of convention: the Celsius and Fahrenheit degrees we are familiar with. In the Celsius scale, the temperatures between the boiling and freezing points of water are divided into 100 equal parts, from 0 to 100, whereas in the Fahrenheit scale they are divided into 180 equal parts, from 32 to 212.

The third scale, Kelvin, is based on a physical principle. The absolute zero, or 0 degrees Kelvin, is the point at which all thermal zigzag motions of atoms and molecules come to a freezing stop. It is one of those never-never limits in science, the speed of light being another one, that we can get very, very close to it, but can never really get there. Things have been cooled down to 0.1 degree Kelvin, 0.0002 degree Kelvin, or even 0.0000001 degree Kelvin, but never really down to the magic of absolute zero.

This limit is extrapolated by a line of data approaching it. Absolute zero is currently pegged at -273.15 degrees Celsius. A temperature of -60 degrees Fahrenheit corresponds to only -51 degrees Celsius. Some interesting and relevant temperatures are marked in FIG. 10–1. For brevity, the three scales—K, C, and F for Kelvin, Celsius, and Fahrenheit—are marked approximately, dropping decimals.

Nitrogen gas liquefies at -409 degrees Fahrenheit, and liquid nitrogen is much cheaper than liquid helium, the two standard refrigerants in superconductor researches. The temperature of interstellar space, the void of nothingness, is actually 3 degrees Kelvin, or -454 degrees Fahrenheit. It is the remnant of the primordial Big Bang. The old and new superconductors—that is, the thresholds at which they were detected to show the properties of superconducting—are plotted against the absolute temperatures in FIG. 10–2.

The initial discoveries in 1911 involved pure elements such as mercury and indium, both of which require cooling by rather expensive liquid helium at temperatures between a few and 10 degrees

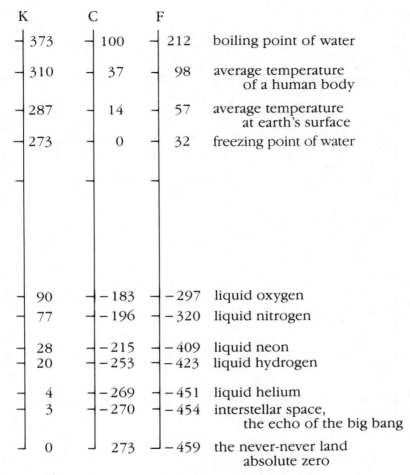

K	C	F	
373	100	212	boiling point of water
310	37	98	average temperature of a human body
287	14	57	average temperature at earth's surface
273	0	32	freezing point of water
90	−183	−297	liquid oxygen
77	−196	−320	liquid nitrogen
28	−215	−409	liquid neon
20	−253	−423	liquid hydrogen
4	−269	−451	liquid helium
3	−270	−454	interstellar space, the echo of the big bang
0	273	−459	the never-never land absolute zero

10−1 *The three temperature scales*

Kelvin. Among all pure elements, niobium shows the highest temperature for superconducting and this has led to developments of about a dozen compounds containing niobium that show superconductivity up to about 30 degrees Kelvin. These compounds include niobium carbide, niobium gallium, niobium germanium, and niobium titanium, some of which are formed into superconducting wires to produce strong but compact magnetic fields for medical purposes. These two groups, pure elements and niobium compounds, had been the only known members of superconductors until 1986.

In 1986, a totally new class of materials—ceramic compounds, which are rather poor conductors of electricity under a normal set of

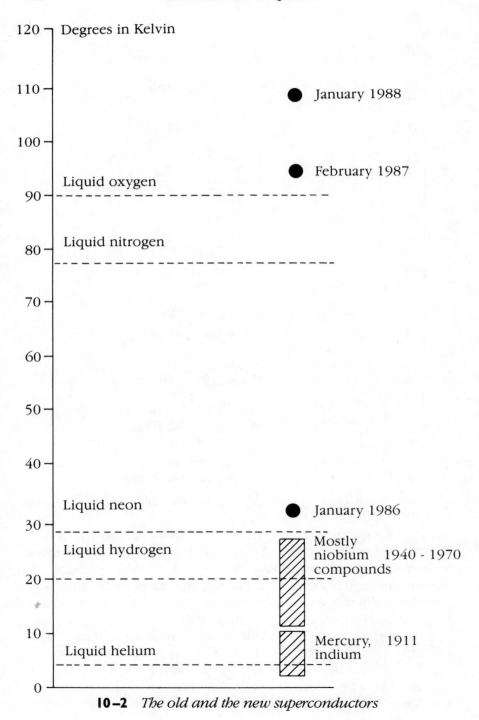

10-2   *The old and the new superconductors*

conditions—were discovered not only to superconduct, but to do so at temperatures higher than anyone had ever suspected possible. This new breed is referred to as *high critical-temperature superconductors;* the name was shortened in the public sector to high-temperature superconductors (HTS). A latest copper-oxide material, which was discovered in January 1988, shows superconductivity at a high 110 degrees Kelvin. Some high temperature it is: − 320 degrees Fahrenheit. HTSs are the focus of attention worldwide, but any return for an investment into a start-up research and development company is likely to be absent from any corporate quarterly reports for some time to come.

Two telltale signs of superconductivity

We have used the word *superconductivity* several times already, and perhaps it is time to define what it is. We can begin by stating what it is not. In chapter 6 we differentiated three groups of materials—insulators, semiconductors, and good conductors—by enormous gaps in their respective free electron densities: 10^2, 10^{12} and 10^{22} free electrons per cubic centimeters, respectively. A superconductor is most emphatically not a super conductor, having a free electron density higher than 10^{22}! The name *high-temperature superconductor* sounds like some dream stuff to be used for heating purposes, such as melting snow and ice from sidewalks, bridges, and airport runways! Well, so much for what it is not.

The ways in which a material conducts electricity or resists the flow of it are clearly two complementary aspects. An insulator has a low conductivity and a high resistivity, and it is just the opposite for a good conductor. The conductivity of a material, whether it is a conductor or an insulator, varies with temperature because the rate of collisions for electrons with the interior lattice arrangement of atoms depends on their thermal motions; less collision at a lower temperature meaning less resistance and more conductivity for the flow of electrons. Generally, you expect the conductivity of a material to improve with the lowering of temperatures, but most materials retain some resistance. For some materials, the resistivity drops to zero when cooled to a sufficiently low temperature, and this is superconductivity.

The great discovery of 1911 was that when certain pure elements, notably mercury and indium, were cooled down to an incredibly cold temperature with the help of liquid helium as refrigerant, it was observed that these materials abruptly lost all resistance to the flow of an electrical current—no resistance, no loss of energy due to collisions at the atomic scale. A current can flow forever, even after being disconnected from a battery or any other power source. Something happens at these frigid temperatures to the lattice structures so that apparently streak after streak of micro-highways open up for electrons to move across without running into a single atom or each other—like an ice hockey puck sliding along effortlessly over a frictionless frozen surface.

The *threshold,* the temperature at which any resistance to an electric current vanishes, is called the *critical temperature.* Not every material becomes superconducting under any temperature, but each material that does has its own characteristic critical temperature. It is a curious fact that all good conductors, except aluminum, never become superconducting; their resistivity approaches some very low value, but not zero. At temperatures as low as 0.05 degree Kelvin, copper and gold retain some residue resistivity. Aluminum, however, buckles under and begins to superconduct at about 1.2 degrees Kelvin. Generally speaking, those materials that are good conductors at the normal range of temperatures are very poor superconductors, and this is strong testimony to the lack of relationship between superconductivity and a high free-electron density.

FIGURE 10–3 shows some typical dependence of resistivity on temperatures for two superconductors and two *nonsuperconductors,* regular conductors that never become superconducting. For materials A and B, when they hit their respective critical temperatures, denoted by t_A and t_B, their resistance drops straight down to zero. For materials C and D, their resistances approach some value, low but not zero. Critical temperatures for some materials are listed in TABLE 10–1.

In addition to the sudden vanishing of resistance, there is another unique property of superconductors: their expulsion of any trace of magnetic field from their interiors. This property leads to the widely photographed effect in which a superconductor pushes off any small magnet that comes near it. A superconducting material is immersed in a pool of liquid helium in a glass container and when a small piece of magnet is dropped onto it, the balance between the gravitational pull and the magnetic repulsion keeps the small magnet floating over it—a *magnetic levitation* as it is called.

This magnetic effect, discovered in 1933 and known as the *Meissner effect,* occurs independently of whether a magnetic field is ap-

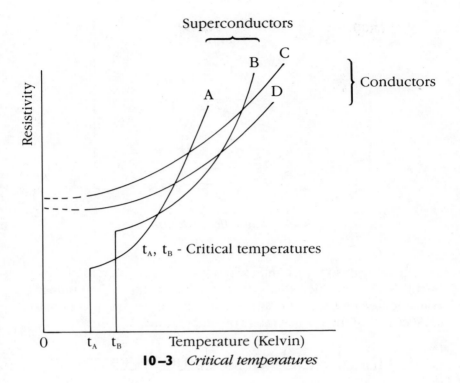

10–3 *Critical temperatures*

TABLE 10–1	*Critical temperatures*	
Aluminum (Al)	1.2 K	− 457 F
Lanthanum (La)	4.9 K	− 450 F
Niobium (Nb)	9.3 K	− 442 F
Thallium (Tl)	2.4 K	− 455 F
Mercury (Hg)	4.2 K	− 451 F
Niobium-germanium	23.0 K	− 418 F

plied either before or after reaching the critical temperature. You cool it down and introduce a small magnet in its vicinity, or keep the setup inside a magnetic field and then cool it. It doesn't matter. Once a material reaches its superconducting state, any magnetic field is exorcised from its interior.

From what we know already, we can say that a superconductivity is not so much a property of a new class of materials—as in insulators, semiconductors, and conductors—as it is a new fourth state of matter, as in gaseous, liquid, solid, and superconducting states. It has

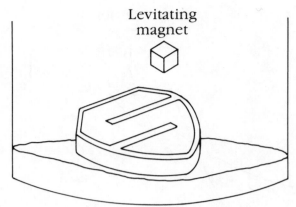

10–4 *Magnetic levitation*

more to do with some yet unknown ways by which interatomic bonds establish very specific lattice structures than with just the number of available free electrons. Ceramics, after all, are very poor conductors of electricity at the normal range of temperatures.

Critical fields and critical currents

A superconducting state, once reached by extensive cooling, turns out to be a very fragile state. It can be lost very easily. It is extremely sensitive to two other external elements, in addition to temperature. If you take the advantage of having a zero resistance and try to pass too much electrical current, the superconductivity suddenly disappears even though the material is still kept at temperatures well below its critical temperature. The amount of current that turns off superconductivity is appropriately called the *critical current*. Critical indeed. What good is a superconductor if we cannot pass enough current for it to carry out some application?

This is a very crucial point. The discovery of new ceramic stuffs is a scientific breakthrough. A real industrial breakthrough will come when we can either discover or fabricate a wonder stuff that can carry a lot of current and still remain superconducting. This we have not yet been able to achieve.

The second critical aspect has to do with the amount of the external magnetic field, which is also capable of suddenly killing off the superconductivity. If a superconducting material is placed within a magnet above a certain strength, it loses its superconductivity. The strength of the external magnetic field at which this happens is called the *critical field.*

Three conditions will kill the superconductivity: temperatures higher than the critical temperature, a magnetic field stronger than the critical field, and an amount of current larger than the critical current. Superconductivity is an extremely fussy and sensitive property indeed! The sensitivity to a magnetic field can be exploited in many technological applications involving a detection of and variations in magnetic fields. A sensitive detection device for submarines is one such application.

The curve in FIG. 10–5 shows a typical interrelationship between the critical temperature of a superconducting material and the strength of the external magnetic field in which it is immersed. The critical temperature is lowered in the presence of a relatively weak magnetic field; that is, a magnetic field impedes a material from reaching its superconducting state. It has to be cooled down further.

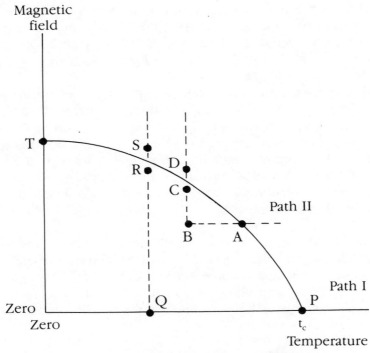

10–5 *Lowering of the critical temperature in a magnetic field*

Point P is the natural critical temperature in the absence of any field, and point A represents the lowered value of it in the presence of some magnetic field. The triangular area bounded by the curve is the superconducting region. With the magnetic field stronger than the value represented by T, the material can never reach its superconducting status.

Let us follow through two paths, marked I and II in FIG. 10–6. Path I represents a situation in which a material is first cooled down then is placed inside a magnetic field, such as being placed in between the north and south poles of some desktop magnet. The material is cooled down to its critical temperature, point P, and it remains in its superconducting status while being cooled further down to a lower temperature, point Q, at which point it is placed inside a magnetic field. As long as the strength of the field is weaker than the value represented by point R, it will continue to superconduct. An increase of the strength of the field from R to S will, however, suddenly destroy the superconductivity.

The situation is similar along path II, except that the superconductivity sets in at a lower temperature, point A, due to the presence of some magnetic field. It is cooled to point B and, maintaining the temperature, the strength of the field is increased. It is OK up to strength C, but it loses suddenly all its superconductivity as soon as the field strength is increased from C to D.

This sensitivity provides us with a remarkable mechanism. Suppose we maintain a superconducting material just below the temperature-field curve, such as points C or R. The electric current that it is superconducting—a supercurrent, so to speak—can be turned on and off just by increasing, by a small amount, the strength of a magnetic field. I am sure that you caught those magic words: on and off! We have a switch! A superconducting switch, no less, and it has nothing to do with doped or, for that matter, undoped semiconductors.

Any mechanism or device that can switch a current on and off, if it works fast, is reliable, can be miniaturized, and can be inexpensively fabricated, would become a prince of a candidate for uses in microchips. There are a whole lot of companies in the world who are quietly pursuing research development in the area of just this kind of superconducting superchips.

AT&T Archives

10–6 *A magnetic cube is placed on a disc made of superconducting material and liquid nitrogen is poured into the container. As the disc is cooled down to below its critical
temperature, below 90 degrees Kelvin, or about minus 298
degrees Fahrenheit, the superconductivity sets in and the magnetic repulsion lifts the cube into its levitation. In the case of
magnetically levitated trains, the cube is the wheel and disc is
the track*

Some conventional applications

As far as the new high-temperature superconductors are concerned, no full-fledged commercial application has yet been developed. Every industrialized country is pouring money into its research and development. One of the most immediate applications will come in the area of superconducting superfast superchips, but beyond that the superconducting high-tech is just only beginning. Whether the new breed of HTS can deliver the kind of industrial revolution as has been reported in media hypes, comparable to what the invention of transistors did for microelectronics, is an open question. The bet is heavily in favor of it. All conventional applications are those utilizing the old superconductors, the kind that have to be refrigerated by liquid helium. Let us have a brief look at some of these conventional applications.

We have already mentioned the magnetic levitation and the extreme sensitivity of superconductors to an external magnetic field. The sensitivity of the critical field is exploited in detecting slight variations in magnetic fields and this property leads to practical applications in geological explorations, detection of mining deposits, as well as a delicate submarine detection device.

A prototype of a magnetically levitated train, the so-called "maglev" train constructed in Japan, operates on the magnetic properties of superconductors. Superconducting magnets are installed to the underside of the train and their motion along the metal track induces magnetic fields on the track, providing sufficient repulsive forces to lift off the train. Ideally, tracks could be built up from superconducting materials, but the cost would be prohibitive. The train travels at a speed up to 300 miles per hour, a frictionless *whoosh*.

Almost all conventional applications are confined in the area of passing currents without resistance. Consider the case of power transmission. For the same amount of power delivered for all our household needs, the higher the voltage, the lower the current through which it can be transmitted. The lower the current, the less resistance an electric current encounters and the less power is lost during the transmission. With superconducting power lines, buried into the ground, there is zero resistance and no power loss and we can transmit as much a current as the wires can bear. Even taking into consideration the cost of maintaining the cold temperatures, most calculations put the estimate of cost savings at no less than 70% to 80%.

The most widespread application of superconductors to date is in superconducting electromagnets, using wires fabricated out of niobium compounds. These wires can carry, on the average, electrical currents 30 to 40 times greater than copper wires, and all without loss of power or heating. Without a heating problem, these niobium compound wires can be coiled up and packed densely within a relatively small volume, thereby providing a very powerful electromagnet that is small in its dimensions.

Thousands of such magnets have been in use at several high-energy particle accelerators in the world—the so-called *atom smashers*. The proposed superconducting super collider, the SSC, calls for about 10,000 superconducting magnets to be designed and built. A much smaller superconducting magnet provides the magnetic field for the advanced medical imaging technique called *magnetic resonance imaging* (MRI), which we will discuss in chapter 12.

There are other areas of applications as well, such as electric motors, electric generators, cars, ships, power storage, and so on, but it gets a little expensive to do all these things using the conventional superconductors, which must be cooled by liquid helium. The new HTS requires cooling by liquid nitrogen—much cheaper and much more readily available. If only we can turn these new materials into practically usable wires. This is one of the most immediate challenges, and at the moment we will just have to wait it out.

11

Nucleons, quarks, and all that

LET US NOW SWITCH GEARS, leave the world of atoms behind and swoosh down deeper into yet another layer of the ultimate constituents of matter; the world inside atomic nuclei, an incredibly small microcosmo of protons and neutrons, and further down, quarks.

The last time we had a skirting encounter with nuclei was back in chapter 2, in which we mentioned that a carbon nucleus is as small compared to its atom as a basketball is compared to a commercial airport, the proportion of sizes being equal to about the ratio of 60,000:1. In an atomic scale where sizes are measured in terms of the sizes of the smallest electron orbit, in so many angstroms, a nucleus can be and is considered, for all practical purposes, to be a sizeless, pointlike particle. It serves two fundamental roles: it is responsible for almost all of the atomic mass, and, with its concentrated positive charges, it is the force center for orbiting electrons.

The brave new world of new forces

From sizes of atoms to the size of the known part of the universe, and for everything in between, two basic forces—the electromagnetic and gravitational—account for all the properties of matter. First, electrons combine with nuclei to form atoms, and atoms coalesce into molecules, all with the help of the electromagnetic force. Molecules condense into liquids and solids, and when these things become massive enough the gravitational force exerts itself, not only on a falling apple but also on a collapsing galaxy.

The two forces share some common characteristics as well. The strength of both decreases rapidly as the square of distances, and both have an infinity range: that is, there is no specific point in distance beyond which either force falls off in its strength by a marked degree. With many common properties and fundamental relevance, it is no wonder that Albert Einstein spent a major portion of his life in an attempt to find a common explanation for the two forces, in vain.

When we get down to the scale of atomic nuclei where sizes are measured in so many fermis, one fermi being 10^{-15} meter, we encounter a whole slew of new subnuclear species: protons, neutrons, mesons, and so on. At still farther down the scale, *nucleons,* as protons and neutrons are collectively called, are believed to be made up from yet smaller constituents called *quarks.* So what appeared as a point-like atomic nucleus turns out actually to have a rather complicated "molecular" structure. Much more significant than encountering new particles, however, is the revelation that when we step into the interior of nuclei, we come face to face with two new forces of nature hitherto unseen and unfelt outside nuclei. These new forces operate only within distances comparable to nuclear sizes and once just beyond their ranges, their strengths drop to zero. It is no wonder that they were not detected until 50 years ago.

One of the two forces is what is commonly called the nuclear force. It is what holds protons and neutrons together to form a nucleus and is called the *strong nuclear force.* As the name implies, this force is very strong. Its strength is about 100 times stronger than the electromagnetic force; in fact, it is the strongest force known. Protons and neutrons are held to each other much tighter than electrons are to nuclei, and this is the source of nuclear energy, the release of which is much more powerful than that of electromagnetic energy. The second new force is what is referred to as the *weak nuclear force,* the least understood of all known forces.

In spite of much progress and some giant strides made over the years, we do not understand these two nuclear forces as well as we do the electromagnetic force. The relative strengths and ranges of the four basic forces are listed in TABLE 11-1.

Extreme disparities among the four basic forces of nature are all too obvious. As stated, the strong nuclear force is operative only within the range of about one fermi, but look at the range of the weak nuclear force. For all practical purpose—whatever *practical* means at

TABLE 11-1 *Four basic forces in the universe*

	Relative strength	Range
Strong nuclear force	100	10^{-15} meters
Electromagnetic force	1	infinite
Weak nuclear force	10^{-11}	10^{-18} meters
Gravitational force	10^{-38}	infinite

such small scales—the weak force has a zero range; that is, you don't really feel it until you are right on top of its source! Disparities among the relative strengths are worse—*relative* in the sense that other forces are compared to the electromagnetic force.

The relative insignificance of the gravitational force explains why this force is completely left out of any consideration when we are discussing atoms, nuclei, and other particles. Due to the tiny amounts of mass these particles carry, the gravitational force plays no part, and we speak of only the remaining three forces, the weak nuclear force being strong enough to matter.

The hierarchy of things inside things is sketched in FIG. 11-1 in which a clear demarcation line is drawn between the world of atoms and that of nuclei. Above this line only one force, the electromagnetic force, accounts for all the physical properties of atoms and molecules. Below the line it is the interplay among three disparate forces that is responsible for all the properties of the parties involved. This tends to complicate things a little, not to mention that we do not fully understand the two nuclear forces yet. The fact that the gravitational force exists but is to be totally ignored is gently indicated by a dotted line. *Subnuclear physics* refers to everything below the demarcation line. It certainly is the study of protons, neutrons, quarks, and all that, but more importantly it is the study of the interplay among three of the basic forces of nature, two of which we have never met before. In the following sections, we will have brief encounters with these forces and meet up with quarks, which embody all three of them.

Gamma emission: "xasers" and "gasers"

Radii of atomic nuclei, as best can be determined experimentally by shooting probing particles toward them, range from 1 fermi for a

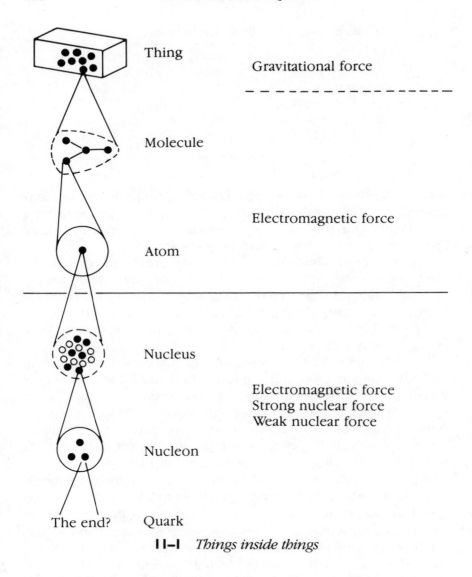

11–1 *Things inside things*

hydrogen nucleus, alias a proton, to about 6 to 7 fermis for a uranium nucleus, 1 fermi being equal to one billionth (10^{-9}) of one millionth (10^{-6}) of a meter—one nanomicron if you like. This range of sizes tells us right away that nuclei are tightly packed balls of neutrons and protons.

The simplicity of this reasoning will surprise you. Let us say that a nucleus is a spherical aggregate of nucleons tightly packed to be touching each other, as shown in FIG. 11-2—something like a completely filled bubble gum dispenser minus the plastic container. The volume of a nucleus will be proportional to the number of nucleons packed tightly inside it, and, as every child knows, the volume of a sphere is proportional to the cube of the radius, the radius multiplied three times. So if this picture is to make sense, we should expect a relation to hold, where the cube of radius of a nucleus is roughly equal to the number of nucleons making it up. This relation holds remarkably well. There is no problem with the smallest nucleus, a proton. One nucleon, and the cube of 1 is 1, and the radius of a proton is indeed one fermi. An einsteinium nucleus is made up from 254 nucleons, 99 protons, and 155 neutrons, and its radius is about 6.6 fermis, the cube of which is about 287. Not bad for an age-old formula from ancient geometry! You would be surprised at the simplicity of some calculations that go on in the stratosphere of high-powered physics.

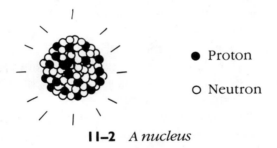

● Proton

○ Neutron

11–2 *A nucleus*

The picture of a nucleus as a tightly packed bunch of protons and neutrons immediately leads to a deduction of utmost importance concerning the nature of the strong nuclear force. Because protons carry the same positive electrical charges, they exert an appreciable amount of mutually repulsive forces. These repulsions are extremely strong at such short distances as protons find themselves in inside a nucleus, the force being proportional to the inverse square of distances. The shorter the distance, the more powerful the force. Yet these protons do not fly apart, but stay glued not only to each other but also to electrically neutral neutrons. Not only must there exist a strong and entirely new force, but this new force must be attractive, and is independent of electrical charges. Since we have not detected any trace of such a new force outside of nuclei, it must be of a very limited short range. Well, this is the strong nuclear force.

Originally, the strong nuclear force was called the *nuclear force* and at the same time it was referred to as the *strong interaction* among professionals. The name *strong nuclear force* seems to satisfy everyone and is becoming a standard terminology. We will say more about this force in the next section, but for now, let us consider the more familiar electromagnetic effects.

Atomic nuclei exhibit discrete energy levels in much the same qualitative manner as atoms. Magnitudes of energies involved are far greater than for atoms—about 1,000 times greater on the average. It takes a lot more energy to elevate a nucleus to one of its higher states, and by the same token when a downward transition is made to a lower level, the energy released is enormous. If that released energy should be in the form of a photon, it has to correspond to a hard x-ray or a gamma ray. This is what is behind the potency of a *xaser,* an x-ray laser, or even a *gaser.*

There are two basic differences between the cases of atoms and atomic nuclei as to the nature of forces that are responsible for energy level structure. Whereas there is a unique center of force inside an atom, the nucleus itself, there is a complete democracy inside a nucleus. Protons and neutrons, roughly of the same mass, are equal partners in making up a nucleus—they all sort of hang around each other and stick to nearby nucleons. There is no "nuclear sun." Furthermore, whereas the energy levels of atoms are the results of only one operating force, the electromagnetic force, three distinct forces combine to contribute in defining energy levels—the electromagnetic, the strong nuclear and the weak nuclear. These energy levels are occupied by nucleons.

A nucleus can be elevated to a higher level by the absorption of a proper amount of energy and return to its ground state by emitting a proper amount of energy, just as the absorption and emission by atoms, but the energy absorbed or emitted is not entirely of electromagnetic origin. Nuclei can absorb and emit quanta of the strong as well as weak nuclear forces, and the emission and absorption of highly energetic photons by nuclei are hence only a part of their activities.

Nuclei can emit three different forms of energy, not surprisingly because there are three forces operative inside. Historically these three processes have been called the *alpha, beta,* and *gamma emissions.* They were originally named after the first three letters of the Greek alphabet when their exact nature was not known. They are usually called *decays,* rather than emissions, such as beta decays and gamma decays. The gamma emission is the straightforward emission of photons, albeit with greater energy than those of atomic photons. Much greater energy means much higher frequency, and this places the nuclear photons in the range of hard x-rays and gamma rays.

In order to produce a nuclear photon laser, we have to repeat the steps discussed in chapters 8 and 9, except in a much greater energy scale. First, a bunch of nuclei have to be excited to a higher level, either by an absorption of a high-energy gamma ray photon or by a similar mechanism involving the strong nuclear force, a small-scale nuclear explosion, say. If enough nuclei can be elevated to their respective metastable levels . . . well, you know the rest. We don't have to do a thing. Sooner or later, one gamma ray photon will be emitted, and it will trigger a chain reaction of cascading events of stimulated emission of x-ray or gamma ray photons. We have an immensely energetic beam of high-energy lasers. This is the Star War program, or the Strategic Defense Initiative. These x-ray laser beams are supposed to be powerful enough to burn right through the metals of intercontinental ballistic missiles.

There are some curious aspects to all this. The x-ray laser beam powerful enough to melt the metals forming the body of a missile must first be reflected off some mirrors and be focused by some lenses, just as in the case of ordinary optics. The laser beams can actually be reflected by some of these specially designed reflectors, which raises an interesting question: What if the exterior of ICBMs are made from, or coated by, these laser-reflecting materials?

The strong nuclear force

Being the strongest force in nature, the strong nuclear force would have been the dominant force affecting everything in the universe if its strength had not been kept in check by a severe limitation on its range of operation. Within the scale of atomic nuclei, the force holds protons and neutrons glued to each other. As the number of protons increases, however, the mutual repulsion among the protons begins to counterbalance the glue. There is an intricate interplay among three different forces, and this makes things a little more complicated and a little difficult to understand. In most cases, naturally occurring nuclei having equal numbers of protons and neutrons are generally stable, and as we go up to heavier nuclei they tend to include more neutrons than protons. Some well-known nuclei and their nucleon contents are given in TABLE 11-2.

TABLE 11–2
Nucleon contents of some nuclei

Nucleus	Number of protons	Number of neutrons
Hydrogen	1	0
Helium	2	2
Lithium	3	4
Carbon	6	6
Nitrogen	7	7
Oxygen	8	8
Silicon	14	14
Copper	29	34
Niobium	41	52
Uranium	92	146

A helium nucleus consisting of two protons and two neutrons is also known as an *alpha particle.* (Historically, it is the other way around. One of the three nuclear radioactivities was called an *alpha emission,* or alpha decay, and later it was determined that the emitted object was a helium nucleus.) The preponderance of neutrons over protons becomes evident starting from a copper nucleus, and it is 146 over 92 for a uranium nucleus.

The strong nuclear and electromagnetic forces treat two camps of a nucleus quite differently. As far as the strong nuclear force is concerned, there is no difference between a proton and a neutron because the force has nothing to do with electrical charges. It is "charge-blind." Each nucleus has a different number of total nucleons; however, the total is divided between protons and neutrons and is a distinct nucleus. A nucleus with six protons and six neutrons is distinct from one with six protons and seven neutrons.

The electromagnetic force, on the other hand, does not care much about neutrons, for an obvious reason, but is dependent on the number of protons inside a nucleus. Since the atomic structures as well as all physical and chemical properties of matter larger than atoms are completely determined by the amount of electrical charges—that is, the number of protons and electrons—we are slightly prejudiced in favor of classifying nuclei according to the number of protons, instead of the total number of protons and neutrons.

Nuclei with the same number of protons but different number of neutrons are grouped into what is called *isotopes*. Members of an isotope group hence have the same atomic structure and the same chemical properties.

Every element from hydrogen up to curium, with 96 protons, is associated with several isotopes. Some have only 2 isotopes while others have as many as 14 isotopes. As shown in TABLE 11-3, the isotopes other than the most abundant ones are either rare or not stable. The naturally occurring stable members, such as carbon-12 and hydrogen, or should we say hydrogen-1, are the most abundant, followed by a small minority such as carbon-13 and deuterium, or hydrogen-2. Other isotopes have a definite lifetime—some short and others long. Look at the disparity in relative lifetime between carbon-11 and carbon-14! What a difference three more neutrons can make!

Most isotopes are, in fact, not stable. Due to the counterbalancing interplay of push and pull between the electromagnetic and strong nuclear forces, these isotopes tend to shed energies and transform themselves into a more stable arrangement of protons and neutrons. Unlike the case of atoms, however, this shedding of energies takes three different forms: emissions of photons, electrons, and helium nuclei.

A few words about terminologies are in order here. These emissions are called *gamma, beta,* and *alpha decays*. Well, the problem is that the word *decay* conjures up an image of something rotting away. Also, the emission of photons by atoms was never called an atomic decay, and there is no point in calling the same process a gamma "decay," so we will just go ahead and use the word *emission.*

A half-life does not mean one-half of a full life; instead, it means the time period in which exactly one-half of an initial amount of an isotope has been transformed into other isotopes by the emission of electrons and alpha particles. After 12.3 years, 1 pound of tritium would have become one-half of 1 pound of tritium, the other half having transformed into other isotopes. A deutron can form a deuterium atom with one electron, just as a proton forms a hydrogen atom. The code for a deuterium ? It is $1s^1$, remember ?

When an oxygen atom forms two covalent bonds with two deuterium atoms, we have a molecule of heavy water, a D_2O instead of an H_2O. The heavy water looks like, tastes like, and feels like ordinary water, but it is heavier and more expensive. An ordinary water molecule contains 18 nucleons, 10 protons, and 8 neutrons, but a heavy water molecule contains 20 nucleons, 10 protons, and 10 neutrons, and it is hence about 11% heavier. In the oceans of the Earth, deuterium contents are about 0.015%. This works out to about 1 gram of deuterium per 16 gallons of sea water, and that is a lot of heavy water. Had it not been so expensive to extract, you could brew some heavy beer that could be quite filling indeed.

A triton is the prime source of neutrons that we need for nuclear warheads and because its half-life is about 12 years, it has to be replenished every 2 to 3 years to maintain the high neutron yield needed to trigger nuclear explosions. You all heard about how many dollars it would take to upgrade old facilities at a nuclear plant in Savannah, Georgia, the only plant capable of producing tritons in the United States.

TABLE 11–3   *Isotopes of hydrogen and carbon*

Nuclei	Number of protons	Number of neutrons	Relative abundance of half-life
Hydrogen, proton	1	0	99.985%
Deuterium, deutron	1	1	0.015%
Tritium, triton	1	2	12.3 years
Carbon-11	6	5	20 minutes
Carbon-12	6	6	98.89%
Carbon-13	6	7	1.11%
Carbon-14	6	8	5,730 years

When we think of a force, it is customary to have a picture of it, as sketched in the first diagram of FIG. 11-3 in which either two masses or two electric charges are shown to be separated by some discernible distance over which they exert forces on each other. Such a picture, even though very widely used, is valid only when the force being described has a respectable range so that some distance can be shown between two sources. Now, since both the size of a nucleon and the range of the strong nuclear force are about 1 fermi—that is, about one nanomicron—two nucleons would have to be in touch with each other before the strong nuclear force can take effect. Two nucleons separated as in the third diagram of FIG. 11-3 wouldn't even feel the presence of each other.

All these renderings in terms of spheres touching and not touching are, of course, highly idealized and simplified pictures, but they help to keep the relative sizes in a correct perspective. A customary picture such as the first diagram, even though it is liberally used, simply does not exist for the strong and weak nuclear forces. Just think what a picture would have to look like for the weak nuclear force, whose range is even shorter. Two nucleons have to be on top of each other, so the two centers can come within the distance of about one-thousandth of a fermi before the weak force comes into play.

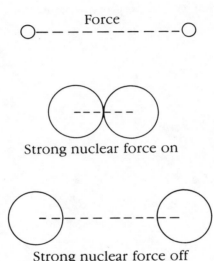

11–3 *The strong nuclear force and its range*

This short range of the strong nuclear force, together with the long-range repulsion of electromagnetic forces among protons, is the key factor for nuclear instability. Suppose we have a nucleus with 81 protons and 124 neutrons. Any one of the protons feels the collective repulsive force from 80 other protons, all trying to push it out. Suppose that the proton in question finds only one other nucleon, a proton or a neutron (its nearest neighbor within the nucleus), within the range of its strong nuclear force. Well, we can subtract two numbers, a strong attraction by a factor of 100 and a repulsion by a factor of 80, and realize that the proton in question is actually hanging in there rather precariously. To exaggerate a bit, a disturbance as little as a gentle sneeze might just be able to pry that proton loose!

An alpha particle, alias a helium nucleus with two protons and two neutrons, turns out to be an extremely tightly glued system, and many heavier nuclei, reaching just such a point of instability, would split and spit out an alpha particle. This is the alpha emission, as shown in FIG. 11-4. When a large nucleus such as uranium-238 splits into two roughly the same size fragments, it is called a *nuclear fission.* This instability resulting out of the interplay between the electromagnetic and the strong nuclear forces is what limits the size of nuclei. We can only place so many protons in a nucleus before it begins to split all by itself. Without this limitation, it would have been possible to have an atom that we can weigh on a bathroom scale!

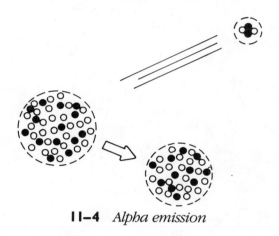

11–4 *Alpha emission*

The weak nuclear force

The subject matter for this and the next section, the weak nuclear force and quarks, is not directly related to any of the high technologies of today, but is an essential part of the subnuclear world, so we will have brief discussions on them. The weak nuclear force is the second new force that operates within the size range of nuclei.

A few superlatives are called for when discussing this force. It is not only the weakest in strength and shortest in range, but is also the strangest and the least understood of all forces in nature. Its strength is ten-trillionths that of the electromagnetic force, and it does not play any role in holding nucleons together, but it is an essential part of the dynamics of nucleons. Because of its extreme short range, all parties that participate in the weak nuclear force must come so close as to be right on top of each other, and the force manifests itself in transforming neutrons to protons, and vice versa.

Under certain circumstances—that is, when it is energetically possible within the energy level structure of a nucleus—the weak nuclear force transforms a neutron into a proton, creating and emitting an electron in the process. Sometimes the role of neutron and proton is reversed, and a proton is transformed into a neutron, creating and emitting a positron in the process. A *positron* is exactly like an electron except that it carries an opposite, positive, charge. It is, in fact, a bit of antimatter, anti- to an electron. These emissions of electrons or positrons by the weak nuclear force are called *beta decays,* or beta emissions, and provide a prime source of positrons for many applications including a modern medical tomography called the *positron emission tomography* (PET). PET has been developed in the eighties and is still in the process of being refined.

Whenever a positron, an antielectron, meets up with an electron, the pair goes into a suicidal explosion in which each gives up everything it has, not only any motion energy but the ultimate source of its own mass, even its own identity. As the pair disappears its energy is turned into two photons coming off in opposite directions. Matter and antimatter collide, disintegrate, and turn into two flashes of light. This is called an *annihilation.*

In PET, a solution containing nuclei capable of emitting positrons is injected into an area to be tomographed, and as photons start flying out of the annihilation of the emitted positrons with plenty of electrons all around in a tissue, the intensity is computer-calibrated to give us a picture, a tomographic image.

A typical electron emission involves the transformation of a carbon-14 nucleus into that of a nitrogen-14 with an emission of an electron, with a packet of missing energy also shooting out in the process. There are many other nuclei participating in the process brought on by a neutron changing into a proton, as shown in FIG. 11-5. All four parties involved—neutron, proton, electron, as well as the missing energy—must all be within a short range for this process to take place. The figure shows a slightly exaggerated separation between neutron and proton, just to show that there is something else underneath. The emitted electron is created at the moment of the transformation. In the second diagram, the complementary situation of the positron emission is shown.

The missing energy has been a source of great debates, soul-searchings, and speculations. It has neither a mass nor an electrical charge, and by the conventional standards of gravitational and electrical forces, it is tantamount to a nothing! It isn't a photon either.

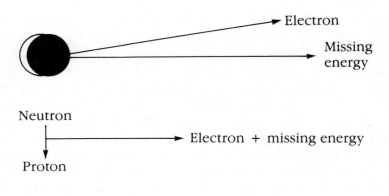

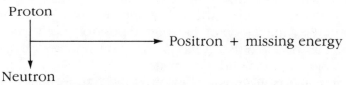

11–5　*The electron and positron emission*

Furthermore, it shows no response to the strong nuclear force. A truly nothing that nevertheless responds to and takes part in the weak nuclear force and it alone. This missing packet of energy, which is almost nothing, turns out to be a particle, called a *neutrino,* a baby neutron. The idea that this is an actual particle was not accepted even by the community of physicists until well into the late fifties, and grudgingly even then.

That is not the end of the story. The missing energy associated with the emission of a positron, an antielectron, is classified, as a *neutrino,* whereas the missing energy associated with the emission of an electron is classified as an *antineutrino.* In case you are getting worried, or antiworried, about all this, let me assure you that we won't get into this subject. Believe me, you won't be missing much by not getting all tangled up in the business of deciding which almost nothing should be anti- to which almost nothing.

The fact that neutrinos have no mass has been brought into question lately, and many experiments have been conducted to determine whether the neutrino mass is truly zero or has some small value, however tiny. These experiments are called the *neutrino oscillation experiments,* and so far they all confirm zero mass. Any potential possibility in which all the masses of neutrinos in the universe, if they had mass, would have a critical influence on the rate of the expansion of the universe has, for all practical purposes, disappeared. In spite of much hoopla about recent attempts to unify various forces, the weak nuclear force and the nature of neutrinos remain two of the least understood topics in modern physics.

Assortment of quarks

In the world of atoms, we had basically two particles and one force to contend with: a proton and an electron, and the electromagnetic force. When we step down to the smaller world of nuclei, we run into two additional forces, as well as two more players. The strong and weak nuclear forces introduce us to a neutron and a neutrino. A *neutron* is a neutral partner to a proton, while a *neutrino* plays that role with respect to an electron. This quartet, or more precisely, a pair of doublets, serves to define, differentiate, and be the sources for the three forces, as shown in TABLE 11-4. The word *lepton* is derived from the Greek *lepto-*, meaning fine, slender, light, and it refers to electrons and neutrinos grouped together the same way the word *nucleon* does with protons and neutrons. Actually, the word *lepton* also means a small coin of ancient Greece.

TABLE 11–4 *Nucleons and leptons*

Twosome	Foursome	strong nuclear	electromagnetic	weak nuclear
		Forces they participate in		
nucleon	proton	yes	yes	yes
	neutron	yes	no	yes
lepton	electron	no	yes	yes
	neutrino	no	no	yes

This picture of a quartet, a doublet of nucleons and a doublet of leptons, was to undergo another fundamental change in its basic makeup. Relentless investigations into the nature of the two new nuclear forces resulted, by the early sixties, in an impressive accumulation of circumstantial evidence, all indicating that nucleons themselves are made up of yet another layer of smaller constituents. These smaller objects making up protons and neutrons were named *quarks* by their proposer Murray Gell-Mann in 1962. Leptons, on the other hand, remain basic to this day.

Two different quarks would make up two different nucleons and in the early days of quarks they were originally denoted as *p* and *n,* that is, a p-type quark and an n-type quark. Needless to say, this created a needless confusion and the names were promptly changed to *u* and *d* for an up quark and a down quark. The art of choosing names became very whimsical around this time, and the names for quarks did not escape it.

I will let you in on a little secret folklore here and tell you how *p* and *n* became *u* and *d.* You type *p* and *n* on a piece of paper, turn it upside down, and you have *d* and *u,* to be called a down and an up!

The quark content of the two nucleons is shown in TABLE 11-5. A deutron nucleus with one proton and one neutron now appears as a complex molecule of up and down quarks, as shown in FIG. 11-6. I will leave it as the only exercise in the entire book for you to draw a picture of a uranium nucleus with 92 protons and 146 neutrons, showing all the quarks, the ups as well as downs. Since a proton is made up out of two up quarks and one down quark, while a neutron consists of one up quark and two down quarks, the electrical charges of quarks turn out to be $+\frac{2}{3}$ and $-\frac{1}{3}$ of the charge of a proton, unlike anything we have encountered before.

Some 30 years since their original theoretical introduction, in fact, not a single quark has ever been detected to date. So many things, indeed, make sense in the subnuclear world when interpreted in terms of the existence of such strange objects, but until and if such

TABLE 11–5
Quark contents of nucleon

Nucleon	Number of up quark	Number of down quark
proton	2	1
neutron	1	2

a quark actually has been observed to exist, they will remain under a fate of being only theoretical constructs.

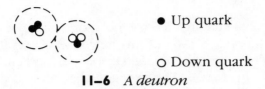

● Up quark

o Down quark

11–6 *A deutron*

The up and down quarks now replace protons and neutrons in the makeup of a fundamental quartet, as given in TABLE 11-4, and we arrive at the current, cutting edge, the state-of-the-art understanding of a fundamental quartet: a doublet of quarks and a doublet of leptons. Quarks form nucleons, nucleons form nuclei, nuclei capture electrons to form atoms, atoms coalesce into molecules, molecules make larger and larger molecules, some molecules get twisted up in the form of a double helix, proteins form tissues, some other molecules in the meantime form solids, some rocks, and others chips, still others some superconductors even, and everybody lives happily thereafter. That, at least, is the tip of the apex of the tower of human knowledge as it stands today. You know, and I know, things are not exactly going to be all that completely understood by us.

The basic quartet is summarized in TABLE 11-6, but already we have to expand our horizons beyond this neat picture of a universal single quartet. Things are never that simple.

First of all, there are more quartets than one! We will not get too much into this here, but we think that there are at least two more quartets, heavier than the first set, but having exactly the same properties with respect to the three forces—in other words, two sets of heavier clones. In the exact order of up, down, electron, and electron neutrino, the second and third quartets have heavier quarks and heavier leptons named as *charm, strange, muon,* and *muon neutrino* for the second set, and *top, bottom, tau,* and *tau neutrino* for the third set. I am certain that you feel relieved to know that we won't be getting into this topic.

You noticed that neutrinos are tagged as electron or muon neutrinos so that we can distinguish one from the other. OK, so we have three quartets: six leptons, and six quarks. The first member of the third set, a top quark, is on a shaky ground, but overlooking that we would have a set of twelve basic building blocks, except for one more thing. It was proposed that quarks carry a new type of charge, not electrical—a tri-valued charge all its own—and the forces among these three different charges are what the strong nuclear force was all about.

I was fortunate enough to have been one of the original discoverers with Y. Nambu in 1964 of this new type of tri-valued nuclear charge, which has come to be called color—red, green, and blue for the three different charges! The strong nuclear force is thus replaced by the color force, the chromo force, among quarks, and each one of the six quarks come in these three charges: red up, green up, blue up, green down, red charm, and so on. Now there are 18 different quarks and 6 leptons—the proliferation never seems to stop.

The name *color* is again just a whimsical choice and it has nothing to do with frequencies in the optical spectrum, nor does it have anything to do with a Sony trinitron or an RGB color display monitor.

TABLE 11–6   *Quarks and leptons*

		\	*Forces they participate in*	/
Twosome	Foursome	strong nuclear	electromagnetic	weak nuclear
quarks	up	yes	yes	yes
	down	yes	yes	yes
leptons	electron	no	yes	yes
	neutrino	no	no	yes

12

Nuclear technologies

THE EMISSION OF GAMMA PHOTONS, electrons, and alpha nuclei is collectively called *nuclear radioactivity.* A radioactive nucleus is an instable nucleus that will go through a transformation by any one or combination of the three emissions.

Radioactive dating

The technique of dating an ancient object, a rock or a fossil, is not exactly one of today's high technologies. It has been around for a while, but as an application of what we discussed in the last chapter, we will have a brief mention of it. Naturally occurring radioactive isotopes, those nuclei that are found to exist naturally but sooner or later transform themselves into other nuclei, have a wide range of half-lives, from a matter of a few seconds to a few billion years. Under the right conditions or a reasonable assumption, they can be employed to date very old objects. As mentioned before *half-life* does not mean one-half of an entire lifetime of some isotope but rather is the time it takes for one-half of an amount to have transformed out while the other half still remains. Suppose we have a radioactive isotope, A, which transform into a stable isotope, B, with a half-life of ten years. If you started out with 512 pounds of A in 1900, you would have just 1 pound of it left by 1990, as shown in TABLE 12-1. Some commonly used radioactive isotopes are listed in TABLE 12-2.

TABLE 12–1

An isotope with a
half-life of ten years

Year	How much left in pounds
1900	512
1910	256
1920	128
1930	64
1940	32
1950	16
1960	8
1970	4
1980	2
1990	1
2000	½

TABLE 12–2

Half-lives of some radioactive isotopes

Isotopes	Number of protons	Number of neutrons	Half-life (years)
Cobalt 60	27	33	5
Strontium 90	38	52	29
Radium 226	88	138	1,600
Carbon 14	6	8	5,730
Uranium 238	92	146	4.5 billion

The ages of objects of biological origin can be determined by using one of the carbon isotopes, since all living things continuously take in carbon dioxide. The ratio of the natural abundance of carbon-14 to the stable carbon-12 is known to be about one part in a billion. Nuclear transformations that occur at the edge of our atmosphere keep replenishing carbon-14 so that this ratio remains fairly constant on Earth, meaning that one part in a billion of carbon dioxide contains the carbon-14 atoms.

Plants consume carbon dioxide and are consumed in turn by humans, animals, and any other plant consuming living things. This way, all living things maintain this ratio of carbon-14 over carbon-12, as long as they are alive and consuming plants. Over a long span of time after death, the amount of carbon-14 atoms, due to its slow half-life of 5,730 years, slowly gets depleted, providing us with a technique of determining the ages of many things after death by monitoring the remaining amounts of radioactive carbons. The ages of mummies and other artifacts of ancient civilizations are determined this way. Normally, this technique of radiocarbon dating is reliable up to about nine half-lives of carbon-14, or up to about 50,000 years.

For dating geological objects, meteorites, moon rocks, and so on, isotopes with much longer half-lives are used. Our knowledge of the age of Earth, about 5 billion years old, or the earliest evidence of humans, about 2 million years old, was gained through this technique. The technique itself we learned only in the last 50 years since the discovery of the neutron.

Diagnostic MRI

A medical diagnostic tool called the *CAT scan* is the second most widely used imaging technique, second only to x-ray pictures. X-rays are used in the CAT scan, the computer-assisted tomography, in the same manner as in the ordinary x-ray pictures. A beam of x-ray photons traverses a sample tissue, but the images are computer processed to give an instant video display.

Less familiar and known is a relatively recent imaging technique, still being refined, based on an entirely different mechanism, which employs photons of such a low energy that it hardly disturbs cells at all. It is a straightforward application of a physical phenomenon called *nuclear magnetic resonance* (NMR), but in medicine it is known as *magnetic resonance imaging* (MRI). The technique combines our knowledge of the magnetic properties of protons and other larger nuclei, the quantum processes of absorption and emission of photons, and the technology of superconducting magnets. It is a quantum-nuclear-superconducting application. Developed for the first time in the 1980s, it is fast becoming a standard medical tool, the whole setup costing about 5 million 1988 dollars.

Magnetic resonance is a befuddling name for a relatively simple mechanism, and we can understand easily how it works by first considering an analogous situation using one small and one large magnet on a tabletop. Suppose we have a small bar magnet that is pinned at the center, but otherwise free to rotate. In the absence of any magnetic field, it will be aligned along any direction. Suppose now that the small magnet on a tray is placed between the poles of a large and

strong magnet, as shown in FIG. 12-1. It doesn't take a genius to see
which of the two configurations, marked A and B in the figure, is the
stable one. As soon as you turn on the large magnet, assumed here to
be an electromagnet, the small bar magnet will spin itself into configuration A, each end tightly pulled by the opposite poles of the large
magnet. You can rotate the bar magnet with a pencil or a small stick
slightly off that alignment, but you would have to hold a pencil firmly
to keep it there. The second you let go of it, the bar magnet will spin
right back into configuration A, proving it to be the stable one. If you
wish to rotate the bar magnet into configuration B, north to north and
south to south, you would have to do some hard and tricky work with
a stick. As soon as you remove the stick, the bar magnet will spin back
to configuration A with a vengeance, exerting force to anything that
gets in its way of rotation. All this is quite obvious, but you have real-
ized by now that we are describing, under a slightly different circum-
stance, an absorption and emission of energy, just like that described
in chapter 8.

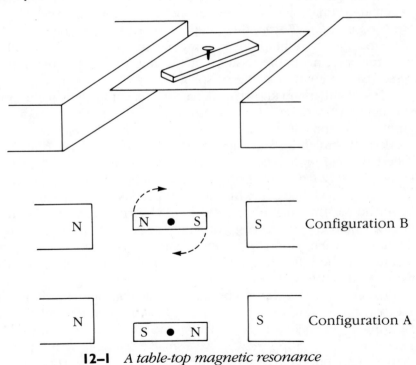

12–1 *A table-top magnetic resonance*

Configuration A is the magnetic counterpart of a stable ground state. You would have to pump in energy—rotating the small magnet with a pencil against the pull of the larger magnet—for the system to go into a higher state, configuration B. Left alone, however, the system immediately returns to the ground state, shedding its excess energy. We have a magnetic counterpart of the absorption and emission of energy, as indicated in FIG. 12-2.

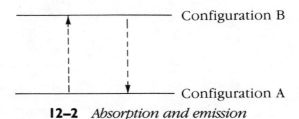

Configuration B

Configuration A

12–2 *Absorption and emission*

Back in chapter 3 we mentioned that a ring of electrical current is what defines the basic units of a magnetic dipole, which is the physical source of the magnetism. Any rotation of a charged object thus acts like a small bar of a magnet (FIG. 3-2). The flow of electrons around an orbit is one such example of a ring of current. There are more examples. Electrons as well as protons rotate constantly about their own axes, at least that is our mental visualization of the situation, sort of like our own planet and others all rotating about their own axes. These are all rings of currents.

Magnetic effects of an electron are too small to be of use in this connection, but the magnetic effect of the rotating, or spinning, proton is large enough to be useful in conjunction with the absorption and emission of energies that we have just described as magnetic resonance. Each proton can be thought of as a tiny magnet, and in this manner our bodies contain a countless number of tiny magnets: every single proton in our body. It turns out that hydrogen atoms are best suited for this mechanism and, as shown in TABLE 5-2, hydrogen comprises 10% of our body by mass, so you might say that we are full of magnets. See FIG. 12-3.

The technique of MRI is now more or less self-evident. In the absence of any strong magnetic field, all these tiny magnets, hydrogen nuclei, are pointing in every which way. As the portion of our body to be scanned is placed inside a strong magnetic field, usually generated

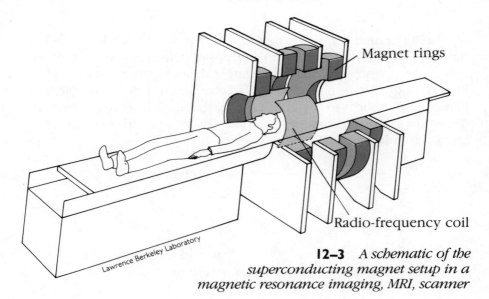

Magnet rings

Radio-frequency coil

Lawrence Berkeley Laboratory

12–3 *A schematic of the superconducting magnet setup in a magnetic resonance imaging, MRI, scanner*

by an expensive superconducting electromagnet (kept in refrigeration by liquid helium because it is still built out of the niobium compound wires), all the protons, acting as tiny magnets, align themselves as in configuration A of FIG. 12-1. At this point, a radio-frequency (rf) source much lower in energy than a microwave keeps radiating energy into our body, providing photons with varying energies as the frequencies are varied over a certain range. The tiny magnets absorb these rf photons and jump up to a higher level, configuration B of FIG. 12-1, and almost immediately drop back to the stable level, emitting a set of rf photons. Out of these low-energy photons an instant image can be constructed with the aid of an on-line computer. See FIG. 12-4.

This MRI technique provides excellent tomographic images, and the whole technology is still going through stages of refinements and improvements. It is relatively a new area of high-tech medicine. Positron emission tomography (PET) that we mentioned in the last chapter is the other example. See FIG. 12-5.

Superconducting super collider

A superconducting super collider is certainly a mouthful of a name, a super name so to speak. This behemoth of a scientific laboratory, which is scheduled to be built on a barren landscape about 25 miles outside Dallas, Texas, at the projected cost of between \$5 billion and

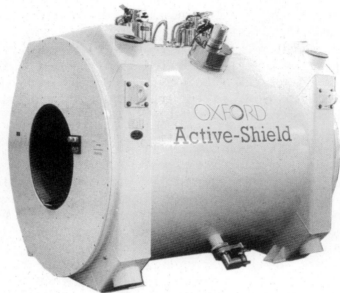

12–4 *A photo of a superconducting magnet assembly used in MRI. The central hollow is the space into which a patient is placed for scanning. The whole thing is cooled by liquid helium*

Oxford Superconducting Technology

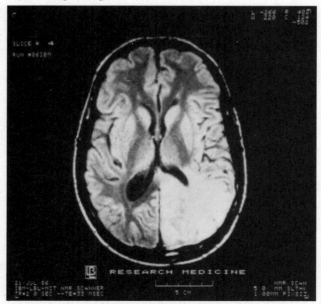

In this MRI image, the white area at lower right reveals the presence of a brain tumor.

$6 billion, is easily one of the most ambitious scientific projects to be undertaken in the world.

The superconducting super collider (SSC) is not, however, the only such gigantic proposal. There is another almost identical proposal on the other side of the Atlantic Ocean, which is called the *large hadron collider* (LHC), to be installed into an already excavated ring of tunnel in Geneva, Switzerland, by a scientific consortium of European countries.

Which will come into being first is an open question. During a lengthy period of public debates concerning the SSC, many questions have been raised and discussed about the whys, whats, and hows of it, but most of the time the science of it has been almost drowned out by the very enormity of it, as well as its economic as well as environmental impacts. We will touch on some of the salient scientific points concerning the SSC in this section.

Both SSC and LHC, representing the newest generation of ultra high-energy particle accelerators, trace their origin to a hand-held contraption called a *cyclotron,* first invented in 1930 by Earnest Lawrence and Stanley Livingston. First of all, what is a particle accelerator and why do we need it ? Well, simply put, it is exactly what it says. In spite of its enormity in scope, its astronomical budget, and the glamorous results it leads to, a particle accelerator is somewhat like a huge bridge in that it is a great big thing that is somewhat crude and has one purpose: to accelerate a particle such as an electron or a proton to a speed coming very close to the speed of light.

When it comes to experimenting with various aspects of the electromagnetic force, we can easily do so in terms of human-sized apparatus assembled on a tabletop—wires, meters, batteries, motors, and so on. Clearly we cannot expect to be able to conduct investigations into the nature of the strong and weak nuclear forces on a tabletop. Because of the severe limitations imposed on their ranges, the only way we can learn about them is by shooting a tiny probe, usually an electron or a proton, deep into the world of nuclei and penetrate into the innards of a nucleus. This is easier said than done, however. If you shoot a jet of hydrogen gas against a side of a common brick, the gas hits the surface, gets deflected, and scatters into air. Penetrate the brick ? No way!

Now a probe must first break through the surface, find its way through a maze of molecules, zero in on one particular molecule, cut through all the electron clouds making up the outer layer of the molecule, find an atom and zero in on it, pierce through several layers of atomic shells, and finally go straight to its nucleus, crack it open, and touch a proton! Now, that takes an enormous amount of energy and power for that little probe.

The next question is how do we go about building up such an enormous energy to a little fellow such as an electron or a proton? We do so by repeatedly accelerating the particle in steps of judiciously applied pulses of forces. In a simpler language, we kick that poor little fellow hundreds of thousands of times with jolts of electricity until it starts flying at speeds comparable to that of light.

The fact in which the amount of energy a particle gains due to its motion depends on its speed is of paramount importance, but this relationship is not as widely known as it should be. It is surprising especially since this relationship is one of the most important contributions of Albert Einstein to modern physics. The energy increases extremely gradually even as the speed approaches 50% of the speed of light, but as it approaches the limit, the speed of light, the energy starts to climb practically on a vertical line. This critical factor was shown in FIG. 12-6 and some typical values are tabulated in TABLE 12-3.

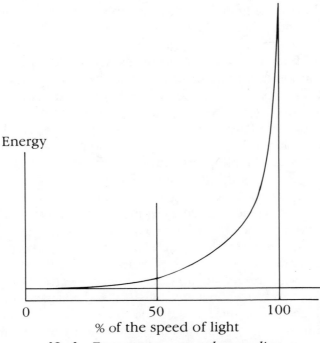

12–6 *Energy versus speed according to the theory of relativity*

TABLE 12–3
How energy depends on speed

Increments of energy	at these speeds as % of the speed of light
1	near zero
1.09	40%
1.15	50%
1.25	60%
1.40	70%
1.67	80%
2.73	90%
10	99.5%
100	99.995%
1,000	99.99995%
10,000	99.9999995%

Accelerating a particle to a speed that is a tiny, tiny fraction more can increase its energy by a factor of 100 or 1,000 in a very narrow range of speed close to the limit. It is to achieve this fractional gain in speed that millions and billions of dollars are needed.

Of the two forces we can employ to accelerate a particle, the gravitational force is too weak to be of any help, especially in view of the totally insignificant masses of particles involved. You can drop a bottle full of protons from the top of the World Trade Center all day, but I assure you that you won't be getting much energy out of them when they hit the sidewalk!

So we are left with only one force, the good old electromagnetic force. The electrical force gives a kick to a charged particle and the magnetic force bends the direction of its motion. As discussed in chapter 3, the electricity and magnetism are two different aspects of the same electromagnetic force, even though, from time to time for the sake of convenience, we talk about them as being either electrical or magnetic forces. They do a marvelous job. Magnets keep a beam of charged particles, electrons or protons, in a circular path and as these particles go around and around, pulses of electrical force make them go faster and faster. In a nutshell, that is what a particle accelerator is all about—simple, crude, and huge—just like those big bridges with large spans, from a small contraption a few feet across to a giant atom smasher several miles in diameter.

The most widely adopted design that is especially well suited for large-scale construction is what is called a *synchrotron,* a design first developed in the early sixties, deriving from a much earlier version originally called a *cyclotron.* A top view of a typical layout is shown in the simplified schematic picture in FIG. 12-7. A battery of electromagnets, as many as a few hundred in each link, keeps the beam of charged particles in a circular motion. Several gaps are incorporated at several locations along the *ring,* as the whole circular assembly is called, and it is here that the particles get the jolting kick of an electrical acceleration each time they jump over the gap. As they go around and around, hundreds of thousands of times, they are accelerated in little increments at a time until they gain the highest speed possible within the capability of an accelerator.

Through the 1960s and 1970s a synchrotron was designed to accelerate one type of particle, mostly protons, and these so-called proton synchrotrons (PSs) would aim the powerful beam of protons onto a stationary target, 1 inch by 1 inch square, of a thin coating of lithium or heavy water. Scientists would then study the shower of all sorts of bits of matter created upon impact.

There are four major such PSs operating in the world today: one each in Europe and the USSR and two in the U.S., one in Brookhaven,

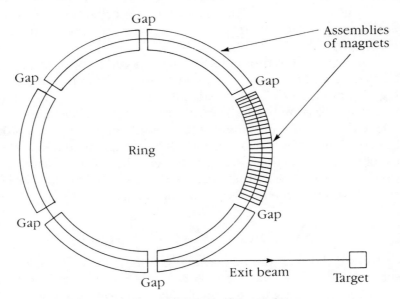

12–7 *A typical layout of a synchrotron*

Long Island (FIG. 12-8), and one at the Fermi National Accelerator Laboratory in Illinois (FIG. 12-9). The ring at the Fermi lab has a circumference of about 4 miles.

In the eighties, accelerator technology advanced further and it became possible to accelerate, simultaneously, two separate beams in opposite directions, contained in separate beam tubes, all in a single synchrotron. After each beam has attained its maximum speed and highest possible energy, they are guided by an array of powerful magnets to come to a head-on collision at a tremendous energy at preselected locations in the ring. You have two circular tracks separated by a short distance, one train running on one track clockwise and another running counterclockwise on the other track. When each is running at full speed you throw a switch and have them come smashing into each other. This is a collider, a synchrotron in which two oppositely traveling beams collide in midflight.

There are two proton colliders in the world: one at the Fermi lab and the other in Geneva, Switzerland. These colliders have already been using superconducting electromagnets and hence are qualified to be called *superconducting colliders*.

When does a collider get promoted to a super collider? Well, when it is really big. When is a market a supermarket, or for that matter a hypermarket? The superconducting super collider is thus a very large (super) synchrotron capable of accelerating two separate beams in opposite directions (collider) employing superconducting electromagnets.

The ring of the proposed SSC is supposed to be about 53 miles all around, and that is a quantum jump in size from the largest existing ring of only 4 miles around, the ring at the Fermi lab. Suppose you drive one complete loop around the beltway of Washington, D.C., I-95 and I-495, and imagine that all along the median, about 10 feet underground, is hidden a continuous tunnel about 10 feet in diameter of the cross section—that comes very close to the picture of the ring of the SSC. The collisions inside the SSC would create, or recreate, a facsimile of the Big Bang in a micro-micro scale.

A cold nuclear fusion?

Just as we did with the name *laser,* examining one letter at a time spelling backwards, the name, *cold nuclear fusion,* can be studied one word at a time starting from the end. A *fusion* means literally two

12–8   *Inside the tunnel at the synchrotron at Brookhaven National Laboratory, showing the massive links of magnets forming a section of the main ring*

12-9 *An aerial view of the Fermi National Accelerator Laboratory. The largest circle is the main accelerator. Three experimental lines extend at a tangent from the main ring. The 16-story twin-towered building is seen at the base of the experimental lines* Fermilab

or more things getting fused together, being united into one by blending or being melted together. We fuse two pieces of metal with the help of a blow torch. More seriously, a fusion process is customarily associated with nuclear fusion. We rarely, if ever, use the word in the context of atomic physics. Whoever heard of an atomic fusion ? An atomic fusion is more familiarly known as the formation of molecules, the molecular bonds discussed in chapter 5. Two nitrogen atoms "fuse" together to form a nitrogen molecule. No one gets excited about these atomic fusions. A nuclear fusion, on the other hand, is something else, the energy released when two nucleons or nuclei are fused together being enormous. It is what is referred to as *thermonuclear energy.*

In a *nuclear fission,* the splitting up of an unstable or nearly unstable nucleus into two or more fragments, the effect of the repulsive electromagnetic force constantly pushing apart the nucleus is what contributes toward the eventual splitting up. In a nuclear fusion, the same repulsive force puts up one of the fiercest resistances of the nature when we try to bring nuclei close enough together for the attractive strong nuclear force to take over. This is the whole trick of nuclear fusion. If we can overcome the fierce resistance and somehow pressure and press together nucleons and nuclei within the incredibly short range of the strong nuclear force, the fusion process will occur automatically with the accompanying emission of protons, neutrons, gamma-ray, and x-ray photons, as well as tremendous amounts of heat.

The electromagnetic repulsion is, however, a formidable obstacle. It is like trying to walk against a strong headwind inside a wind tunnel! Normally, extremely high temperatures are needed for the fusion process to occur as they do inside the Sun and other similar stars where temperatures are in the range of a few million degrees. In case you noticed that the scale of temperature, Fahrenheit or Kelvin, is not specified. When it is a few million degrees, the differences among the scales are irrelevant. Such high temperatures are difficult to attain on Earth, and despite some 40 years of fusion research, we have not yet arrived at any practical quantity of sustained fusion reaction.

Many known reactions of nuclear fusion involve such nuclei as carbon-12, carbon-13, nitrogen-13, nitrogen-14, nitrogen-15, oxygen-15, as well as lighter nuclei such as helium-4, helium-3, triton, deutron, and proton. (TABLE 11-3 gives the nucleon contents of some of these nuclei.) A helium-3 is very rare, consisting of two protons and one neutron, one neutron less than the stable and abundant helium-4. A triton consisting of one protein and two neutrons is a sort of the flip side of a helium-3.

The most well-known nuclear fusion reactions involving three to four nucleons are listed in FIG. 12-10. They are:

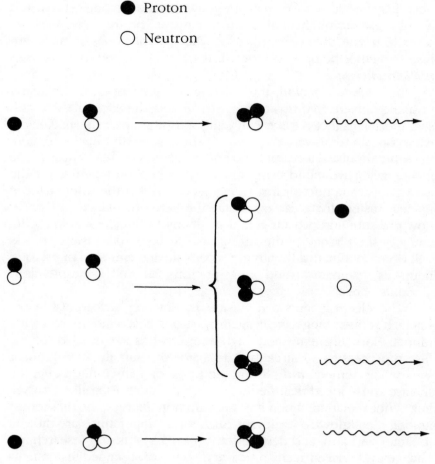

12–10   *Pictures for nuclear fusion*

1. proton	+	deutron	→ helium-3	+	photon
2. deuton	+	deutron	→ triton	+	proton
3. deutron	+	deutron	→ helium-3	+	neutron
4. deutron	+	deutron	→ helium-4	+	photon
5. proton	+	triton	→ helium-4	+	photon

Of these, process number 4 is a rather rare occurrence. Reactions number 2 and 3 are the most predominant ones, each of which has a clear signature of emitting one quite energetic nucleon, be it a proton or a neutron.

Recently we have had some exciting reports of just such a fusion reaction, except that it was claimed to have been achieved under room temperature, the so-called *cold nuclear fusion*. By immersing a rod of palladium metal, or other metals such as titanium, into a glass jar containing heavy water and passing electrical currents through it connected to a nondescript battery, a deutron-deutron nuclear fusion has been claimed to occur.

Under the influence of an electrical current, deuterium atoms are thought to be pulled in and embedded in the lattice structure of the palladium metal, somewhat like the way in which impurities are embedded into a silicon lattice to produce the doped semiconductors. Deuterium atoms so embedded at lattice points are supposed to be close enough that the strong nuclear force can begin to do its trick. Anyway, that is the most plausible speculation, assuming what has been seen is truly a nuclear fusion.

There are a couple of difficulties with this on both theoretical and experimental grounds. No convincing amount of neutrons have been detected to date, and the best estimate of the distance of separation between atomic sites in a lattice of some of these metals used in the experiments is at least 1,000 to 10,000 times farther apart than the range of the strong nuclear force. Of course, there is always the possibility that we are bumping against something totally new and hitherto unknown. A cold nuclear fusion, if true, would provide humanity with an inexhaustible supply of energy. The relative abundance of deuterium is 0.015%, but that is 0.015% of all the water on Earth, and that is a lot of deuterium indeed.

Glossary

alpha decay A nuclear emission process in which an unstable nucleus emits a high-speed alpha particle and transforms itself into a different nucleus. Despite the name, there is nothing that is rotting away.

alpha particle A helium nucleus comprised of two protons and two neutrons.

angstrom A unit of length particularly suited for the world of atoms, being equal to one-tenth of one-billionth of a meter (10^{-10}), or about four-billionths of an inch. Sizes of atoms range between one and three angstroms.

antiparticle To each particle, there is a corresponding antiparticle, which has the same mass but equal and opposite electric charges. A proton has a negatively charged antiproton and an electron has an antielectron, called a positron. Every member of the basic building blocks of matter, quarks and leptons, is associated with its own antiparticles. An antineutron is made up out of antiquarks and is distinct from a neutron. If an antiproton and a positron were to form an antihydrogen atom, that atom would be distinct from a hydrogen atom even if it were electrically neutral. Antineutrinos are opposite to neutrinos with respect to the weak nuclear force.

beta decay One of the three nuclear emission processes called nuclear radioactivity, in which an unstable nucleus goes through a transformation, emitting an electron and a neutrino. This is the work of the weak nuclear force.

binary code A system of two-unit alphabets in which things are expressed by combinations of them, such as one and zero, yes and no, on and off, or the presence or absence of an electric current.

bit A single unit of a binary code, either one or zero. The term is shortened from *binary digit*.

byte A set of eight bits that forms a single unit of information and data whether it is an alphabet, a number, a punctuation symbol, or a control function such as Enter. Capacity of memory chips is usually given in terms of bits, such as kilobits and megabits, whereas the total capacity of a computer system is usually given in terms of bytes, such as kilobytes, megabytes, or gigabytes.

central processing unit Abbreviated CPU. The "brain" of a computer that executes arithmetic calculations and controls the flow of data. For smaller computers such as personal computers and workstations, the entire CPU can be fabricated on a single chip. Such a CPU-on-a-chip is called a *microprocessor*.

charge Traditionally, the electric charge, the sources of the electromagnetic force. Generally, the sources of other forces, such as color charges for the strong nuclear force.

chip A popular term for an integrated circuit about the size of a baby's fingertip, containing thousands or millions of transistors and other components; also known by megachips and superchips, and so on.

collider An advanced version of a particle accelerator in which two separate beams of particles—electrons and positrons, protons and protons, or protons and antiprotons—are accelerated in opposite directions and brought together for a series of controlled collisions at extremely high energy, creating or recreating a shower of elementary particles.

color charge One of three different source charges of the strong nuclear force: red, green, or blue. Color charges are to the strong nuclear force what electric charges are to the electromagnetic force.

deutron A nucleus consisting of one proton and one neutron. It is the nucleus of a deuterium atom, which forms a heavy water molecule in exactly the same way hydrogen atoms form a water molecule with an oxygen atom. It is one of the prime candidates for a nuclear fusion fuel.

electromagnetic force One of the four basic forces in nature, which include gravity as well as the strong and weak nuclear forces. The force that acts between electric charges. It holds atoms and molecules together, but inside a nucleus its repulsive force works against the nuclear forces.

electromagnetic radiation A wave of electric and magnetic fields that propagates through space at the speed of light, carrying its own energy. Its spectrum covers radio waves, TV waves, microwaves, infrared waves, light, ultraviolet rays, X-rays, and gamma rays. It carries a massless form of radiation energy.

electron An extremely light and negatively charged particle that orbits around a nucleus thus forming an atom. Motion of electrons through a good conducting material is called an electric current. A premier member of the lepton family, electrons can exist independent of atoms.

energy A measure of the capability, endowed by a force applied to it, of matter to do things such as to move, to exert forces to others, or to heat up things. A force sets a matter into motion, giving it a kinetic energy. A force may relocate a heavy object to the top of stairs, giving it a potential energy. At sufficiently high energies, masses can be converted into electromagnetic radiation energy and vice versa.

fermi A unit of length used rather exclusively in nuclear physics, being equal to one-thousandth of one-trillionth, or one-millionth of one-billionth, of a meter (10^{-15} meters). Nuclear sizes range between one and seven fermis. One fermi is equal to 100,000 angstroms. It serves to illustrate the relative proportion of sizes between atoms and nuclei.

field As a concept diagonally opposite to that of a point particle, in principle, any quantity that is defined over an extended region of space. The term is used mostly in connection with the influence of long-range forces spread out over an extended space and time, such as the electric field, magnetic field, and gravitational field.

fission A splitting of a large nucleus into fragments, usually into two smaller nuclei of roughly the same size. A fission is triggered by shooting a high-energy neutron into a sample of fissionable large nucleus such as uranium. It is the source of nuclear energy in a nuclear reactor (controlled) or in a nuclear bomb (uncontrolled).

fusion The opposite of a nuclear fission, in which two light nuclei such as deutrons are brought together close enough to fuse into a single nucleus with a release of a great amount of energy. This is the source of the Sun's energy. Attempts to tame this source of energy here on Earth have been going on for over 30 years with a very limited success.

frequency   The number of cycles per second. For a moving wave, the number of crests passing through a given point per second; for a generator of electrical pulses, the number of on-and-off cycles per second. Frequency is measured in units of hertz. A computer that boasts a speed of 20 megahertz has on-and-off pulses at the rate of 20 million cycles per second.

gamma decay   The third member of nuclear radioactivity, in which a nucleus emits a high-energy radiation corresponding to a gamma ray. The fact that this emission of an electromagnetic radiation is one form of nuclear radioactivity lends itself to an easy confusion between a *radiation* and a *radioactivity.*

gravitational force   The weakest of the four basic forces of nature and easily the most familiar. Even though this purely attractive force dominates the behavior of large objects, in the microworld of atoms and nuclei, it is completely negligible.

high-definition TV   The next generation of advanced color television having more than double the picture lines of today's sets and so capable of producing sharp images of printed picture quality. Not only does it represent a quantum leap in the picture quality, but also it will be an important factor for the chip manufacturing industry due to its great appetite for microchips.

high-temperature superconductor   Abbreviated HTS. Very recently discovered copper-oxide family of materials which attains superconductivity at a relatively high temperature, on or above the boiling of liquid nitrogen. The race is on to develop this new material into commercial applications.

integrated circuit   Also called a chip. An entire electronic circuit, sometimes even a whole computer, whose microminiaturized electronic components, such as transistors, are fabricated on a single piece of a semiconductor material, such as silicon. Typically a rectangle of about 1/2 by 1/2 inch, it is the marvel of today's high technology.

isotope   A nuclei with the same number of protons but different number of neutrons than the element to which it is associated. A deutron and a triton are the heavy isotopes of a proton, the nucleus of a hydrogen atom. Every element has several isotopes.

large-scale integration   Abbreviated LSI. The technique of fabricating a chip with thousands of electronic components. A lot more components than an LSI is a VLSI, a very large-scale integration.

laser   Originally, an acronym for *l*ight *a*mplification by *s*timulated *e*mission of *r*adiation. A beam of light that is of a single frequency, hence of one color, travels in one direction, and most importantly contains all coherent all waves; that is, waves that move and fluctuate in steps—crest-to-crest and trough-to-trough.

lepton   A family of relatively light particles that do not participate in the strong nuclear force, and this separates them from quarks. Most familiar members are the electron and the neutrino, as well as their antiparticles, the positron and the antineutrino.

magnetic levitation   The magnetic repulsion of a superconductor that will lift a magnet placed on it and allow the magnet to float in the air.

maser   A laser that uses a beam of microwave radiation produced by using molecules, instead of a beam of light.

memory chip   A chip whose electronic components form thousands and millions of electronic cells, each storing a single bit of information by being either on or off. A random-access-memory (RAM) chip temporarily stores bits internally during the processing of a program, and loses all memory when turned off (also known as read-and-write memory). A read-only memory (ROM) on the other hand, is a permanent storage for data and instructions.

microcomputer   A computer based on a microprocessor. Includes all personal computers and workstations. Ordinarily any desktop or stand-alone computer is a microcomputer.

micron   A unit of length equal to one-millionth of a meter, a scale suited for the fabrication technology of microchips, in which the separation distance between two adjacent components are in terms of so many microns. A submicron technology will make a much more powerful chip.

microprocessor   A central processing unit built on a single chip. Sometimes called a *computer-on-a-chip.*

nanosecond   Literally one-billionth of a second. It is a common unit of measurement of the operating speed of faster computers. A frequency of one gigahertz corresponds to one cycle of one nanosecond. A fast computer running at the speed of several nanoseconds is common.

neutrino   The lighter member of a lepton family. It is perhaps the strangest of all particles observed to date. It has no electric charge, no color charge, and either zero or very little mass. It zips across the void with the speed of light and comes in three different species: the electron-type, the muon-type, and the tau-type. Just how one is exactly different from the other is still a mystery.

neutron   The neutral partner of a proton that together with the proton forms an atomic nuclei. Protons and neutrons are identical particles as far as the strong nuclear force is concerned, and they are generically referred to as *nucleons*.

optical fiber   Hair-thin fibers made from glass compounds through which light, usually in the form of laser pulses, is transmitted wholly contained inside the tube. A bundle of these fibers forms an optical cable for transmission of on-and-off laser signals.

optoelectronics   As the name suggests, a merging of the laser and microelectronics technologies. It is still relatively a new and young branch of high technology.

particle accelerator   A huge device, often several miles in its dimension, designed to accelerate charged particles such as electrons, protons, positrons, and antiprotons to a very high speed by means of electromagnetic force. Most of them are of circular design—such as synchrotrons, colliders, and super colliders—but some are of a straight-line design such as a linear collider.

photon   A quantum of radiation energy, a "particle" of light, so to speak. The energy of a photon is proportional to the frequency of radiation of which it is a quantum. The energy of a photon of radiation A whose frequency is 1 billion times that of radiation B is a billion times the energy of a photon of radiation B.

proton   The nucleus of a hydrogen atom. Together with an electron, it forms a pair of perhaps the most important and stable members of the particle family.

quantum   An elemental unit of energy, an indivisible packet of energy. The most well-known quantum is that of electromagnetic radiation: the photon.

quantum mechanics   A technical term for the branch of physics dealing with the mathematical formulation of the physics of atoms, nuclei, particles, and their interactions with radiation.

quark   A set of fundamental particles that carry the color charges of the strong nuclear force and that make up other particles, such as protons and neutrons. Quarks participate in all four basic forces of nature. Despite an impressive amount of indirect evidence, so far not a single quark has been detected in isolation. One of the main purposes of the proposed superconducting supercollider is to unravel the deep mysteries surrounding quarks.

radioactivity A generic term for the three nuclear processes of alpha, beta, and gamma decays.

semiconductor 1. Naturally, a solid substance whose ability to conduct electrical currents lies between that of a good metal and an insulator. Semiconductors include silicon, geranium, and gallium arsenide. 2. Natural Semiconductors with impurities introduced. These so-called *doped semiconductors* have improved conductivity and are what is referred to as semiconductors.

strong nuclear force The strongest of the four basic forces of nature. It operates between color charges of quarks, just as the electromagnetic force operates between electrical charges of charged particles. Sometimes also called the color forces—a chromo force, so to speak. The nuclear force between protons and neutrons is seen as a "molecular" force arising out of this color force.

supercomputer A computer that uses a unique processing technology called *parallel processing,* in which a job is automatically divided up and processed concurrently.

superconductor The state of any of a handful of metals and alloys under an extreme condition of very low temperatures whereby it loses all resistance to the flow of currents. These currents can thus flow without any loss of power. Recently discovered high-temperature superconductors are a new breed, being made of ceramics, which are very poor conductors of electricity under normal conditions.

superconducting supercollider A large (super) collider in which all electromagnets used are the superconducting type.

synchrotron A most widely used form of particle accelerator in which charged particles are accelerated in a circular path of a fixed radius, hundreds of thousands of times around. A variation of a synchronism is the *collider.*

transistor A very versatile and reliable semiconductor device that acts as a switch. The invention is heralded in today's high-tech and the age of information.

triton A heavy, rare, and unstable isotope of a proton consisting of a proton and two neutron. Used as the primary source of neutrons to trigger nuclear fission in nuclear warheads atop intercontinental ballistic missiles, it has to be replenished once in a few years.

wavelength The distance between two successive crests in a wave. Like markings on a ruler, the wavelength of a radiation signifies its power of resolution. The shorter the wavelength, the higher the resolution, the higher the frequency, and the higher the energy of its photon.

weak nuclear force The second weakest of the four basic forces of nature. It has the shortest range among the four. All quarks and leptons participate in this very weak force.

workstation A generic name for a powerful microcomputer usually having much more memory and higher speed than personal computers. Primarily used for engineering and scientific calculations, it also has a superb graphics capability.

x-ray laser Similar to a laser except that it produces a beam of x-ray radiation by using atomic nuclei.

Index